Human Nature and World Affairs

About IEA publications

The IEA publishes scores of books, papers, blogs and more each and every year – covering a variety of stimulating and varied topics.

To that end, our publications will, in the main, be categorised under the following headings, from May 2023 onwards.

Hobart Editions
Papers or books likely to make a lasting contribution to free market debate, either on policy matters or on more academic grounds

IEA Foundations
Primers or introductions to free market concepts, ideas and thinkers

IEA Perspectives
Majoring on current policy-focused issues

IEA Briefings
Short, punchy reports on contemporary issues

IEA Special Editions
Spotlighting lecture transcripts, celebrated writers and more

Much of our work is freely available on the IEA website. To access this vast resource, just scan the QR code below – it will take you directly to the IEA's Research home page.

IEA Foundations 1

HUMAN NATURE AND WORLD AFFAIRS

An Introduction to Classical Liberalism
and International Relations Theory

EDWIN VAN DE HAAR

Institute of
Economic Affairs

First published in Great Britain in 2023 by
The Institute of Economic Affairs
2 Lord North Street
Westminster
London SW1P 3LB
in association with London Publishing Partnership Ltd
www.londonpublishingpartnership.co.uk

The mission of the Institute of Economic Affairs is to improve understanding
of the fundamental institutions of a free society by analysing and expounding
the role of markets in solving economic and social problems.

A CIP catalogue record for this book is available from the British Library.

ISBN 978-0-255-36827-8

Many IEA publications are translated into languages other
than English or are reprinted. Permission to translate or to reprint
should be sought from the Director General at the address above.

Typeset in Kepler by T&T Productions Ltd
www.tandtproductions.com

Printed and bound by Hobbs the Printers Ltd

www.carbonbalancedprint.com
CBP2250

To Nigel Ashford

CONTENTS

ABOUT THE AUTHOR

Dr Edwin van de Haar is an independent scholar who specialises in liberal international political theory.

In his academic work he is fascinated by the different views on international relations within liberal political thought, and his aim and motivation for research and writing are to shed light on these differences and to foster an intra-liberal conversation on international affairs.

Dr Van de Haar is the author of *Classical Liberalism and International Relations Theory: Hume, Smith, Mises and Hayek* (Palgrave Macmillan, 2009), *Beloved Yet Unknown: The Political Philosophy of Liberalism* (Aspekt, 2011, in Dutch) and *Degrees of Freedom: Liberal Political Philosophy and Ideology* (Routledge, 2015). He contributed to a number of edited books, among them *The Oxford Handbook of Adam Smith* (2013), *Just and Unjust Military Intervention: European Thinkers from Vitoria to Mill* (2013) and *The Liberal International Theory Tradition in Europe* (2021). His articles on international issues and on liberal thinkers, most notably Hume, Smith, Mises, Hayek and Rand, have been widely published, including in *The Review of International Studies, International Relations, International Politics, Economic Affairs and Contemporary Southeast Asia*. He regularly contributes to (Dutch) public debate.

Dr Van de Haar has been a visiting fellow and lecturer in political theory at John Tomasi's Political Theory Project at Brown University, a lecturer in international relations (IR) and political economy at the Institute of Political Science at Leiden University, and a lecturer in international relations at the European Studies Program at Ateneo de Manila University.

He received his PhD in International Political Theory from Maastricht University in 2008, and also holds masters degrees in international relations from the London School of Economics and Political Science, and in political science from Leiden University.

Dr Van de Haar is a Dutch national and lives in The Hague. He is a member of the Mont Pelerin Society.

For more information, visit www.edwinvandehaar.com.

ACKNOWLEDGEMENTS

Writing this book has been a great pleasure. It gave me the chance to refresh, expand and improve the arguments I put forward in *Classical Liberalism and International Theory: Hume, Smith, Mises and Hayek* (2009) and *Degrees of Freedom: Liberal Political Philosophy and Ideology* (2015). Some new ideas are based on other academic work, while others are specific to this book. As an introduction, its intended audience is broader, which explains the different tone of voice, referencing and writing style compared to the previous books.

Although the phrase 'I am honoured' is often abused, in this case there can really be no doubt. Ever since I met Lord Harris in the autumn of 1996, when he addressed the Hayek Society at the London School of Economics and subsequently invited me to that year's Wincott Lecture, the IEA has been special to me, not least because of its long history of outstanding publishing by so many people whose work I admire. So yes, I am honoured to also make a modest contribution.

I am very grateful to Professor Lord Syed Kamall, former Academic and Research Director of the IEA, for his interest in classical liberalism and international affairs. He invited me to contribute to several IEA outlets, commissioned this book and also completed a thorough edit. His successor,

Professor James Forder, saw the project to an end with enthusiasm, for which I am also grateful. My gratitude as well to all other people involved in the production of this book, as well as the communications on it.

I also thank the two anonymous reviewers, whose suggestions definitely improved the book. Any remaining errors are solely mine, of course.

Floor Schipper has been a great source of love and support. We have been able to create rather ideal circumstances to combine work, study and leisure. This is fantastic, and I thank her for her encouragement and support.

The book is dedicated to Dr Nigel Ashford of the Institute of Humane Studies (IHS) at George Mason University. We met in 2005 at the IHS Graduate Summer Seminar at the University of Virginia in Charlottesville. Nigel has always had a keen interest in the topic of classical liberalism and international relations, and above all he is a very congenial man. Thanks to him, I was able to join several IHS programmes when I wrote my PhD thesis, and ever since, he has never tired of connecting me to people in his wide network who have an interest in IR. Therefore, in the spirit of Nigel's astonishing and valuable work, I hope this book whets the appetite of many students and scholars to start working in the field of classical liberal international political theory.

Cheers, Nigel!

SUMMARY

This book presents a classical liberal theory on international relations. It is not newly developed, but distilled from the writings, and based on the ideas, of major classical liberal thinkers, particularly David Hume, Adam Smith, Ludwig von Mises and F. A. Hayek. Hence, while the elements of the classical liberal theory have always been present, they were also scattered among different places in their work. Their ideas on international relations were never presented comprehensively and this allowed for many erroneous interpretations.

Classical liberalism

The classical liberal theory of international relations is a logical extension of classical liberal ideas about domestic politics. It is quite distinct from other versions of liberalism. The classical liberal theory will be presented through a number of building blocks and related concepts, loosely following the method developed by Michael Freeden (1996). The main ideas of the classical liberal theory of international relations are:

- The starting point is a realistic view of human nature, which values the social nature of individuals but also recognises their propensity to quarrel, fight and use violence. Views on human nature are crucial for international relations because all international politics is about human activities.

- War and violence are regrettable but inevitable features of international affairs. Classical liberals seek to deal with that rather than attempt to get rid of violence and war.
- The (nation) state is the main actor in international relations. Sovereignty should be respected. Only in exceptional circumstances is moderation in the form of federation called for.
- There is no harmony of interests in world politics. States will be able to form an international society, but they will also always face a security dilemma.
- Balance of power is a form of spontaneous order at the international level.
- The just war principles apply, but there is a very limited number of cases that justify military intervention, mostly only to prevent or stop genocide.
- On the one hand, international law and international organisations are useful, and international obligations count and should be kept (*pacta sunt servanda*). On the other hand, international law and organisations can also pose major threats to individual liberty. Therefore, classical liberals only support a limited amount of both.
- Sound economics at the international level means free trade and globalisation, while classical liberals are sceptical of development aid provided by governments and paid for by taxpayers.
- While free trade improves relations between countries, it does not in itself foster peace. It can neither change human nature nor overcome other causes of war.

Differences with other liberal theories of IR

The classical liberal theory of international relations differs substantially from existing liberal international relations theories. The main differences are summarised below.

Liberal IR theories	Classical liberal IR theory
World peace is attainable, in the belief that humans are rational enough to overcome war and conflict.	Conflict and war are perpetual characteristics of international relations, based on a realistic view of human nature.
The nation is seen as a problematic actor in world affairs.	The nation is the primary and a natural actor in international relations.
Balance of power is problematic and a cause of war.	Balance of power acts as a spontaneous ordering mechanism, to a degree fostering international order.
A full catalogue of human rights needs to be defended.	Only classical human rights need to be defended.
Peaceful international relations can be fostered by domestic institutional arrangements, most notably democracy (democratic peace theory).	Sceptical about possibility of domestic arrangements to overcome conflict and war.
Important role for intergovernmental and non-governmental organisations, regimes and international law, which aim to overcome or neutralise the effects of the logic of power politics.	Role of international law, regimes and intergovernmental organisations is important but should be limited and is mostly functional. They can easily become a threat to individual liberty.
International trade is expected to foster peace.	No necessary relation between trade and peace.
Fairly broad support for military intervention, also for democracy promotion.	Military intervention only acceptable in exceptional instances, such as genocides. Democracy promotion hardly ever successful.

Differences with realism

Classical liberalism is not a 'realist' theory in IR. Realism and classical liberalism have a number of ideas in common, such as the central place of the nation state in world politics, the appreciation of the balance of power and the recognition that war is sometimes inevitable. However, classical liberals

- are more positive about the possibility of international order than realists and value moral standards in world politics;
- are more concerned with individual liberty than the interests of the state;
- are less eager to embrace great power management, because the sovereignty of all states should be ensured – international law and organisations should be limited, but cannot be discarded;
- view world politics not as outright anarchy, but as an anarchical society of nation states, since a global system of anarchy is less predictable and less open to classical liberal ideas of just world politics, such as globalisation, free trade, just war rules and limited but effective international law;
- prefer only a small number of functional international organisations, out of principle;
- are sceptical of development aid, which is often granted for political power reasons, but more supportive of individual donations to fund non-state aid and assistance.

Differences with libertarianism

Classical liberals and libertarians share a preference for individual liberty, free trade and globalisation, and are concerned about the expansion of the state in a war. However, classical liberals

- value the place of the nation in international politics, the balance of power and the possibility of peace in international relations;
- are less concerned with stopping 'the American empire';
- do not necessarily support completely open immigration and reject the idea of a positive relationship between trade and peace;
- do not believe in guerrilla warfare or the privatisation of national defence;
- view libertarian thought on international relations as having an isolationist side, especially unilateral isolationism and/or neutrality.

TABLES

1 INTRODUCTION

Pandemics, war in the Ukraine and elsewhere, international trade, Brexit, power struggles between great powers, development aid, international organisations, the European Union, the United Nations Security Council, nuclear and chemical weapons, cyber warfare, border disputes, international environmental and climate policies: these are just a few of the topics covered in the news media, online and offline, every day, all over the globe.

What do classical liberals think about these issues? Not much, it seems. While scholars of Liberal Internationalism and Liberal Institutionalism, as well as scholars of the Realist, Constructivist, Marxist, Critical Theory, Feminist and other schools regularly comment on foreign affairs, not much is heard from classical liberals. There may be the odd academic article or commentary on mostly American current affairs by classical liberal writers, but it often appears that many classical liberals believe that if we just have free trade, globalisation and isolationist foreign policies, all will be fine. For many people, inside and outside the liberal tradition, this disqualifies them as serious partners in debates on international relations.

This is troublesome, because it leads to:

- erroneous portrayals of the ideas of the main classical liberal thinkers;
- statements based on selected, often isolated parts of the works of classical liberal writers, which do not represent their full views;
- erroneous statements about classical liberalism in relation to world politics, including by classical liberals themselves;
- the absence of classical liberalism in academic theories of liberalism and, as a consequence, an incomplete portrayal of the liberal tradition;
- generations of students receiving an education without the faintest idea that classical liberalism not only covers international economics but international relations as well, which may in turn make classical liberal ideas appear to be less relevant since they live in an increasingly globalised world.

This oversight needs to be addressed. This book provides a classical liberal way to look at world politics. It offers a classical liberal lens for the analysis of international affairs. For many people, including academics in the field of international relations theory, this might appear to be novel, as they might not have realised such a classical liberal theory of international relations exists. Yet there is and always has been such a theoretical approach to world affairs, originating in the classical liberal tradition that roughly commenced in the eighteenth century.

Philip Cunliffe is therefore the proverbial exception among academics, acknowledging in his book *The New*

Twenty Years' Crisis (2020) that liberalism in international affairs could perhaps even be traced back to the Dutch Revolt against the Spaniards (1568–1648), but certainly to the 1756–1763 Seven Years' War, when the British claimed to defend liberty against French absolutism, and Hume suggested that the balance of power was a requisite to preserve liberty.

Classical liberals look differently at world politics compared to other liberal thinkers, let alone conservatives, social democrats, Christian democrats, socialists or communists, to name just a few. The differences with the other liberal ideas on international relations will be highlighted in this book, because that is where most confusion occurs, both within the wider liberal tradition and in academia and beyond. For example, some IR textbooks refer to 'classical liberal internationalism', which allegedly lost its relevance after two world wars, but what is really meant here is a classical view of liberal internationalism, not the international dimension of classical liberalism (Jönsson 2018). We will get to that in the third part of the book.

International political theory

This book is not a commentary on international affairs; it is also not about party politics. Rather, it is about political philosophy applied to the international arena. This subject area is called international political theory (Brown and Eckersley 2018). Political philosophers tend to focus mostly on internal (domestic) politics, often overlooking or downplaying the influence of events and ideas beyond the border. At first glance, international relations theorists seem to do a better job of incorporating ideas on international relations from philosophers and thinkers, yet their approach is hardly ever structural. They tend to pick and choose from the history of ideas and seldom attempt a comprehensive analysis of these ideas within the whole intellectual legacy of a thinker. Often, IR theorists call a theory 'liberal' on the basis of two or three ideas they associate with liberalism, such as individualism, capitalism or democracy. This is insufficient.

International political theory tries to be the lynchpin between political philosophy and IR by retrieving and taking further the ideas of thinkers on international affairs. The results are theories of international relations that have a solid base in political theory. This is the basic approach taken in this book, with a central place for the ideas of influential classical liberal thinkers, most notably Hume, Smith, Mises and Hayek. The classical liberal theory of international relations presented here was not invented but is the result of delving into their ideas and works. The author previously explored this classical liberal theory of IR in *Classical Liberalism and International Relations Theory* (Van de Haar 2009). Those ideas are now expanded, refined and improved.

The claim here is to present 'a theory'. To some extent, this is a rather large claim to make. To put things in context, some explanation of theory in IR is needed. There is not just one way to theorise about international relations or to propose a new theory. According to Halliday (1994),

IR theory is the analytical and normative theorisation of interstate relations, transnational relations and the international system itself. IR theory is multifaceted, methodologically and epistemologically pluralist. Within that large space for theorisation, classical liberalism is presented as a world view. Griffiths (2011) defines this as 'a distinctive set of ideas and arguments about international relations, with a discrete set of concerns, sustained and reinforced by a body of causal reasoning about how international relations work'. A world view highlights certain types of issues, actors, goals and types of relationships while ignoring or deemphasising others.

In this text, a set of building blocks will be presented that, when combined, present the classical liberal view on international relations. Together they cover major questions of IR, such as the place of the nation, war, the balance of power, military intervention, international cooperation, immigration, and much more. This classical liberal view on international relations is descriptive and prescriptive at the same time, seeking to answer the main questions: what is the (possible) effect of international relations on individual liberty and how is individual liberty best preserved in a world where international relations is increasingly important?

The answers can never be complete, as the world continues to develop in unforeseen ways. In that sense, every theory is a lens, not a strict guide. Therefore, the classical liberal theory of international relations is necessarily incomplete but its main and timeless features can certainly be presented. It is hoped that there is enough on offer, particularly for students and those interested in classical liberalism, liberalism more broadly and international relations theories.

It is important to note that IR is a rather neglected topic among those people who identify as classical liberals. Hence, they may not always be aware of the theory presented here or may routinely mix a number of liberal ideas, which will be presented in part III. Yet, the author's contention is that they should make the effort to become more consistent.

Lastly, not all parts of the classical liberal theory are set in stone or beyond debate, for example, its view on the European Union or on immigration. Nevertheless, this book provides a theory that is fully consistent with the views of the most important classical liberals both on domestic politics and on international relations.

Structure

The next introductory chapter examines different strands of liberalism. The succeeding chapters are divided into three parts.

Part I presents the main ideas on international relations of four important classical liberals: David Hume, Adam Smith, Ludwig von Mises and F. A. Hayek. This is meant to inform readers about these thinkers' ideas on international relations and to present these ideas in relation to each other. On the occasions (often rare) when students of IR are exposed to the ideas of any of these four thinkers, they are often only taught an incomplete view of their ideas. In line with this, these ideas are also often missing in IR textbooks but also in classical liberal publications.

Part II is concerned with the main building blocks of the classical liberal theory of international relations, which are largely based on the writings of Hume, Smith, Mises and Hayek. It is divided into five chapters, examining:

- *Individuals*: liberalism is the political expression of individualism in both the domestic and international spheres. What is the classical liberal view on human nature and individual rights and how does this relate to international relations?
- *Groups*: Humans are social beings and live in groups. What role do groups play in the international arena?
- *Violence*: individuals and groups sometimes get into conflict. What, if any, is the role of violence in international affairs and how should it be dealt with?
- *Rules*: is there a place for international rules, and if so, what rules?
- *Economics*: how does international relations deal with the topic of economics?

Part II closes with a presentation of the main structure of the classical liberal theory of international relations.

Part III is about IR theory. It presents the current liberal theories of international relations and compares them to the classical liberal theory. A comparison with the major IR theory of realism is provided, as is a comparison between classical liberalism and libertarianism. To bridge the gap to practical politics, some guidelines for a classical liberal foreign policy are also included.

Concluding remarks

This book is intended as an introduction or primer; hopefully it is written in an accessible style. Arguably, every topic discussed in the following pages deserves a book in itself. This text often just scratches the surface. Readers interested in a particular topic will find ample guides to further reading in the extended bibliography.

As is customary, international relations (with small letters), international affairs, world politics and world affairs are used interchangeably, and refer to real-life events. International Relations (IR) in capital letters refers to the academic discipline. And, for every 'she/her' used in this book, readers may also think the reference is to a 'he/his', or any other term one feels comfortable with!

2 LIBERALISMS (AND CONSERVATISM)

Using the term 'classical liberalism' suggests that other forms of liberalism are not 'classical'. Over the years, scholars have written about many forms of liberalism including libertarianism, bleeding-heart liberalism, economic liberalism, political liberalism, social liberalism, high liberalism, minarchism, Objectivism, anarcho-capitalism, etc. Many academic texts mention neoliberalism without even attempting to define it. In international relations theory, you can find neoliberal institutionalism, liberal internationalism, liberal institutionalism, embedded liberalism, and others. The wide range of liberalisms can also be found in other academic subjects. Clearly, this is incomprehensible for both the lay reader and academics. Eamonn Butler has attempted to clarify these liberalisms, most notably in his IEA primers on classical liberalism (Butler 2015) and his book *101 Great Liberal Thinkers* (Butler 2019).

In line with Butler, it is argued here that getting a decent grasp of liberal political thought does not have to be complicated. As a rule of thumb, we should keep one of the perennial questions in political philosophy in mind: what is the just relationship between the individual and the state? In basic terms, there are three liberal answers: the state should have (almost) no role in individual life, the state should have a limited role, or the state should have a fairly large role. The liberal variants that are associated with these answers are libertarianism, classical liberalism and social liberalism, respectively. These three are not completely mutually exclusive, while the thinkers associated with these variants do not always neatly fit the categorisation, certainly not over the course of their whole careers. Still, this basic division into three types of liberalism is as good as any other and has the advantage of offering a simple yet well-argued classification of liberalism (see Van de Haar 2015).

Conceptual approach

The differences between these three liberalisms are also considered by Michael Freeden in his book *Ideologies and Political Theory* (Freeden 1996), where he distinguishes between ideology, political theory and political philosophy. We do not need to go into detail here, yet the reader should note that all three have a common characteristic, namely that they are composed of a set of political ideas, or concepts, that stand in a particular relation to each other, for example, liberty, individualism and natural rights. Although their precise meaning is sometimes contested, the concepts are the building blocks of a political theory or political ideology. They vary in importance; there are core, adjacent and peripheral concepts. The precise relationship between them is called a 'morphology' by Freeden. On the basis of the main writings of major thinkers associated with a tradition, it can be determined which concepts are important and how these concepts stand in relation to each other, and therefore what the content and meaning of a political theory is.

In the case of the liberal tradition, there is a need to distinguish between three different variants, even though each is still part of the larger liberal family. For example, while the concept of liberty is central to all liberalisms, liberty has multiple meanings (is 'contested'). Isaiah Berlin's famous divide between negative and positive liberty is relevant here (Berlin 1969). The former can be defined as 'the freedom from interference by others', the latter as 'the freedom to fully enjoy one's rights and liberties', which often demands some support from the state. Classical liberalism is associated with negative liberty and social liberalism with its positive meaning. This shows in the writings of, for example, F. A. Hayek and John Rawls, respectively. Yet the meaning of negative liberty may be further contested. For example, the protection from interference by others may be taken as absolute or total protection, but many classical liberals do not oppose the compulsion of governments levying taxes on individuals in order to pay for public services. While classical liberals will support the lowest possible taxes, many libertarians (such as Murray Rothbard) may view taxation as an important infringement of individual liberty. To make matters even more complex, libertarianism itself is divided into those who hold an absolute idea of negative liberty (the anarcho-capitalists) and those who allow minimal infringement of property rights to pay for police, external defence and the judiciary (the minarchists, such as Ayn Rand).

To explain the above, a simple framework of liberal concepts, an example of Freeden's morphological framework, is presented in table 1, which is then briefly introduced.

Table 1   A morphology of liberalism

	Classical liberalism	Social liberalism	Libertarianism
Core concepts	Negative freedom, realistic view of human nature, spontaneous order, limited state	Positive freedom, positive view of human nature, social justice as self-development, extended state	Negative freedom, realistic view of human nature, spontaneous order, natural law including strict defence of property rights, no or minimal state
Adjacent concepts	Natural law, rule of law/ constitutionalism	Modern human rights, rule of law and neutral state, social contract (Mill: utilitarianism)	Minarchism: minimal state, rule of law
Peripheral concepts	Social justice, strict defence of property rights	Property rights, spontaneous order	Social justice

Source: Van de Haar (2015).

Classical liberalism

Classical liberalism originated in the eighteenth-century Scottish Enlightenment, especially in the writings of David Hume and Adam Smith (who were influenced by

French Enlightenment thinkers, among others). It is also associated with later thinkers such as Ludwig von Mises, F. A. Hayek, Milton Friedman and James Buchanan. Classical liberalism takes a realistic view of human nature, which means that humans are seen as a mix of rationality and emotion, so they are not guided by reason alone. Individual freedom is the main classical liberal goal and is best preserved by the protection of classical human rights, the rule of law, and reliance on spontaneous ordering processes in society, such as the free market.

However, it should be pointed out that classical liberals do not see humans as just individuals by nature. Classical liberal thinkers, such as Adam Smith, agree with John Donne that 'no man is an island'. Mises (1996) writes about social cooperation, while Ashford (2003), a modern classical liberal, writes about civil society being 'all those voluntary organisations that exist between the individual and the state, such as the family, churches, sports and music clubs, and charities. It is a common mistake to suppose that an individual existing alone can be free.' Classical liberals see humans as social beings.

The classical liberal state is limited, which means it only has to perform or arrange a limited number of public tasks and services. Besides defence, police and judiciary, this includes a minimal amount of welfare arrangements, some environmental regulation, or other public goods that cannot be dealt with through the markets. Classical liberals disagree on the precise size of the state but prefer it to be smaller than social liberals would like and larger than a libertarian state.

Social liberalism

Social liberals, in general, are liberals in the contemporary American sense. In the UK, social liberal thought originated in the nineteenth century. Most notably in the writings of John Stuart Mill and his successors, such as the late-nineteenth-century New Liberals (among them Leonard Hobhouse, Thomas Hill Green, John Hobson and David Ritchie). Since the 1970s, John Rawls and his followers have been the major sources of intellectual inspiration for social liberals worldwide. For social liberals, libertarian and classical liberal ideas allow for a world full of social injustice. Individuals need to have the capacity to develop their talents and should be able to learn skills and get the right knowledge to use their natural talents in the labour market and elsewhere. They also need to be able to fully participate in democratic decision-making processes. Otherwise, to them the idea of liberty is just formal and without much practical meaning. This concern for social justice entails the redistribution of income to ensure widely accessible education and a welfare system (social security, public health) that takes care of the less fortunate. This leads to a much bigger role for the state and higher tax bills than the other two liberalisms deem just. Social liberals do not think the forces of spontaneous order are sufficient to achieve their goals. Their positive view of human nature means they think reason can, in the end, overcome the emotions. This leads to trust in rationally constructed public arrangements, usually via the state, with the goal of individual development, which social liberals see as the real meaning of liberty.

Libertarianism

Libertarianism, like social liberalism, has its origins in the nineteenth century and may be found, for example, in (some of) the writings of Lysander Spooner, Herbert Spencer and William Graham Sumner. Libertarians criticise classical liberals – let alone social liberals – for allowing the state to grow too large. Instead, the strict protection of individual natural rights to life, liberty and property ensures a just society. Significant traces of natural law thinking can also be found in classical liberalism, but they are seen by libertarians as justifying more infringements of property rights. Libertarians favour a system where free people will be able to use their talents and cooperate in strictly voluntary ways. Some, such as Murray Rothbard or Hans-Hermann Hoppe, argue this society can totally rely on spontaneous order for the provision of all necessary services, and therefore they want to abolish the state completely. Others, such as Ayn Rand, think there is a need to publicly organise defence, police and judiciary via the state. Few, if any, libertarians think that there is a need for a centrally organised redistribution of resources, for example, to advance ideas of social justice. Instead, they rely on spontaneous forces and gestures to assist disadvantaged people in society. Many libertarians think it is unjust to keep people inside a particular state if they do not want to; they should be allowed to form new political entities (secession).

Application in IR theory

Freeden (1996) concentrates on domestic politics, but the differences among the three liberal variants are also clearly visible in their views on international affairs. For example, in the role of the nation in individual life and in global politics, there is also the perennial liberal question of whether free trade fosters international peace, or the alleged usefulness of international governmental organisations. This book will mostly focus on the classical liberal theory of international relations. Most current liberal theories in international relations tend towards being socially liberal, as discussed in part III. Libertarian international relations is hardly discussed in IR theory texts, but will also be discussed in the third part of this book.

Differences between liberalism and conservatism

Before turning to the ideas of the four classical liberal thinkers, we should clarify the differences between liberalism and conservatism. There is often some confusion between liberalism and conservatism, especially in Europe, where classical liberal–minded politicians can be found in both liberal and conservative political parties. Also, free market policies are often associated with conservative political parties. Some conservatives in politics embrace liberal ideas, while some liberals defend ideas commonly associated with conservatism, such as the societal value of family life. Since both conservatism and classical liberalism cover a spectrum of views, there is bound to be some overlap between the two.

Therefore, for purposes of comparison, mainstream conservatism is presented here as found in the writings of Edmund Burke, Alexis de Tocqueville, Lord Acton, Michael Oakeshott, as well as in Roger Scruton's *The Meaning of Conservatism* (2001), Robert Nisbet's *Conservatism* (1986) and Russell Kirk's *The Conservative Mind* (1985). American neoconservatism, which was popular at the beginning of the twenty-first century, is not included.

Conservatism's view of human nature is more negative than liberalism's. Conservatives view humans as being capable of doing good but often inclined to do evil. Like classical liberals but unlike social liberals, conservatives do not have high expectations of the power of human reason. The conservative thinker Michael Oakeshott opposed the 'rational illusion' which is fundamental to social engineering, believing that the focus on reason overlooks experience, history and moral virtues (Oakeshott 1962). Scruton adds that the main difference between conservatism and all forms of liberalism is that conservatives do not value individual liberty as the ultimate value of political conduct and political thought. For conservatives, the individual is not unique but formed by social customs and society. Conservatives argue that it is a 'liberal myth' that humans are individuals by nature. Individual freedom is not absolute but must be measured against the possible damage it may cause to the social fabric. Hence, negative freedom is not a goal in itself. So, conservatives support the violation of privacy if this is deemed necessary to reach some higher goal, such as state safety.

Conservatives do not principally object to government interference in society. For them, the individual and society are inseparable and society is an organism where everyone must play their part. Order is important, hence the emphasis on tradition, traditional norms and values, habits and customs. The teaching of the (cardinal) virtues can be of help to address human weaknesses. The government's first task is to provide order, which is legitimate as long as it is done constitutionally. Another core conservative value is the acceptance and demand for authority. People are born unequal and remain so in terms of status and abilities. The preservation of order demands authority and power placed in the hands of small elites that others must obey (which Burke called 'the natural aristocracy'). This idea, or rather 'attitude', is fostered within social institutions such as the family, church, schools, the army, guilds, and so forth. Property, inheritance and family are also positively correlated in the conservative mind not for their value to the individual (as is the case with liberals) but for their value to the social fabric of a country, especially the property of land.

Conservatives do not oppose change, but judge changes against their desire for order and safeguarding society. Undesired change is seen as dangerous, akin to the demolition of society, unnatural, or revolutionary. Conservatives are not defenders of the status quo, but favour slow organic change, not suddenly overturning societal order, which is the result of 'the wisdom of the ages'. This is not unlike Popper's piecemeal social engineering, beloved by Hayek, who applied this to argue against collectivist political change. Liberals are more optimistic than conservatives about the changes brought about by entrepreneurial dynamics, technological change or scientific innovation.

Religion plays a larger role in conservatism than liberalism, especially for social conservatives on issues such as abortion, gay rights, euthanasia, etc. While conservatives do not hesitate to use state power to regulate or prohibit such issues, most liberals claim these issues fall within the individual's private sphere. Many liberals are religious but draw the distinction between the private and public spheres, not wishing to impose their private religious views and practices on others, whereas conservatives tend to bring religious issues into the public sphere.

While conservatives and classical liberals may agree on the importance of free markets, conservatives are more inclined to public interference, for example, to foster or protect national champions. They tend to be a counter-movement: when there is a left majority, they lean to the right, but when there is a right majority they lean to the left. Scruton (2017) believed that 'Liberals seek freedom, socialists equality and conservatives responsibility. And, without responsibility, neither freedom nor equality have any lasting value.'

The main difference between liberalism and conservatism relevant to this book is that most conservatives, especially neoconservatives, are associated with the realist IR theory (as we shall see in chapter 11).

In terms of Freeden's theory, the morphology of conservatism is given in table 2.

Table 2   A morphology of conservatism

Core concepts	Negative view of human nature, organic change, order, groups/family, hierarchy
Adjacent concepts	Active state, free market/spontaneous order, counter-movement, (land) property
Peripheral concepts	Individual rights, freedom

Source: Van de Haar (2015).

PART I: THINKERS[1]

---

1   This part is largely based on Van de Haar (2009). Please refer to that book for more detailed referencing of the primary sources. In addition see Van de Haar (2008, 2013a,b, 2022).

3 SCOTTISH ENLIGHTENMENT: DAVID HUME AND ADAM SMITH

In this first part of the book, we examine the main views on international relations of four great classical liberal thinkers: David Hume, Adam Smith, Ludwig von Mises and F. A. Hayek. This chapter will look at the two Scottish philosophers, while the two Austrian thinkers will be the subject of the next chapter. In part II some of their views will be fleshed out in more detail, to help build the classical liberal theory of international relations.

The focus on these four thinkers is not intended to deny the importance of other classical liberals, such as Milton Friedman or James Buchanan, but to recognise that any attempt to arrive at a classical liberal theory or view of international relations cannot be seriously made without considering their views.

It should also be acknowledged that both Hume and Smith were influenced by previous generations of writers, ranging from the ancient Greek and Roman thinkers to seventeenth- and sixteenth-century writers, such as Bacon and Hobbes, and Scottish Enlightenment thinkers, such as Carmichael, Hutcheson, Shaftesbury and Lord Kames. Where relevant, these other thinkers will be mentioned.

David Hume (1711–76)

While David Hume is admired as a philosopher, it is often overlooked that he also wrote about international affairs and had his own experience as a diplomat. His earliest official international experiences were as an assistant to the Scottish–American soldier and politician General Arthur St Clair in 1746 and 1748, first on a mission to Western France and later on a secret mission to Vienna and Turin. In the mid 1760s, Hume worked at the British Embassy in Paris, initially as personal secretary to the ambassador and later as embassy secretary and *chargé d'affaires*. Finally, he served as undersecretary of state for the Northern Department (which later became part of the UK's Foreign Office) for almost a year between 1767 and 1768. However, he published most of his work on international relations before his posting to Paris.

Hume's starting point on international affairs began with consideration of the individual (also see chapter 5). He contended that the concept of a nation raised positive or negative passions in all people. He saw national pride as the most positive passion, which was caused by direct experiences, such as the beauty of the landscape, and indirect ones, such as the goodness of the produce of the country or the pleasure of the people in the country. It was impossible for individuals to develop a real passion for a foreign country. While individuals valued the same moral qualities in people, whether of the same nationality or from overseas, Hume contended that we feel closer with our countrymen than with foreigners. While these sentiments may appear quite nationalistic to the modern reader, Hume was certainly not a nationalist and remained an internationalist in his writings and in his personal life (Van de Haar 2008). Moderation was key, while extremes on either side were not. All plans depending on a change in human nature were doomed to failure because they were plainly imaginary.

Sovereign states, all of which have a right to national sovereignty, were the central actors in world politics. Intervention was not appropriate, not even when a state's behaviour was improper. Sovereignty was sometimes limited; for example, no country could claim ownership of the sea. Yet under normal circumstances, international order depended on states and their mutual cooperation. Neighbouring countries had a duty to maintain good relations, 'suitable to the nature of that commerce, which they carry on with each other'. The world was not an anarchy, and there was no perpetual struggle for power, but an international society of nations (certainly in Europe), characterised by cooperation, diplomacy and rules, such as the immunity of ambassadors, the principle of the declaration of war, the prohibition of the use of poisoned weapons and the obligation to treat prisoners humanely.

The laws of nations were translations of, and additions to, the domestic laws of nature, which were (1) the stability of possession, (2) its transfer by consent, and (3) the performance of promises. These also applied internationally, had the same benefits and worked the same way. Without respect for property rights, war would be the norm internationally. Without a mutually agreed transfer of property, commerce would not develop. If promises were not kept, alliances or treaties were useless. Hume was influenced by Pufendorf and Grotius in this regard (Harris 2015). His work is perhaps the best expression of the classical liberal 'bottom-up' approach towards international relations.

However, domestic and international politics were not identical. Hume thought that the obligations in the international arena were not as strong. The moral obligations of princes were weaker. This was not meant to ignore treaties or play Machiavellian games. It was more a recognition that international rules were less fixed and that in emergencies, states might decide to dispense with certain rules of justice or the keeping of treaties. This was only to apply in emergency situations and not to become the normal state of affairs in international society. Hume (1998) believed that international law had less force than domestic law but should be respected in normal times.

Hume's best known essay about international relations was on the balance of power, which he praised as a secret in international politics that added to better international management. Its central aim was to prevent domination through the use of force by a large power, by forming a countervailing coalition. It was not a magic concept, but one based on common sense and reasoning. Hume saw it as the duty of statesmen to ensure it worked and believed that a sole focus on domestic affairs was a serious neglect of duty. While the balance of power might result in a status quo of international order, Hume acknowledged that it was also fragile and could occasionally trigger a chain reaction towards war and violence. Despite this risk, he viewed the balance of power as generally beneficial since it could keep even the most powerful empires in check (Hume 1985).

Hume was uneasy about war, although he acknowledged it was a central feature in international relations (Whelan 2004). It should be limited, but hoping for it to disappear was unrealistic. War could and should be used when it was justified, for example, when one country threatened another's freedom. In line with his natural law credentials, Hume also thought that war needed to be just. It needed to have a justifiable cause but even then leaders should be prudent before engaging in war. He saw war for frivolous causes as wrong, since the immediate effects were devastating, often spilled over to other nations, incurred considerable expense and damaged free commerce. Compared to other international law thinkers, he focused more on the negative effects of war.

His views on empires developed over the years. He initially saw European empires as important in advancing knowledge, the arts and industries, as well as increasing levels of commerce. In Britain, he welcomed them for offering opportunities to more people and not just the upper class. Later in life, he became more critical of conquests, confiscation of land and impingements on individual liberty in the colonised lands, although he never called for an end to the British empire. He came to prefer exchange on equal terms and independence, especially in relation to the American colonies. He was an early defender of their independence and corresponded with a number of Founding Fathers. In 1775, he called himself 'an American in my principles' (Hume 1932). He did not believe that Great Britain would suffer a great permanent economic or geopolitical loss from American independence. He conceded that there might be a loss in reputation but saw this as inevitable since he believed that a war against the Americans could not be won.

Hume (1985) was a great defender of international trade, which he saw as enabling economic growth as well as social and cultural development. He thought that commerce, the greatness of a state and the happiness of its inhabitants were related. He also saw overall well-being as empowering the wider public against the elite. For him, trade and commerce could be sources of opulence, grandeur and military achievement, as long as they were accompanied by free government and liberty. Since Hume viewed imports as a sign of opulence, and not a threat to the welfare of a state, he strongly rejected mercantilism as the 'jealousy of trade' and those who argued for a positive balance of trade, i.e. the value of exports exceeding the value of imports. He also saw an interdependence among nations in that the increase of wealth in one nation usually promoted the wealth and commerce of its neighbours.

Hume also argued that foreign trade and international power were related, since a richer society would be able to spend more on its own defence. However, Hume disagreed with the notion that trade inherently promoted peace; he believed that human nature could not be changed by trade. He viewed this as a regrettable effect of trade but believed that the advantages outweighed the disadvantages (Manzer 1996).

Hume was very much a thinker of his time but even today he would be viewed as a moderate thinker, who endorsed the middle way in international politics. For him, a relatively orderly international society of states best promoted individual liberty. His domestic political views and his international political views were balanced and generally consistent with each other. He opposed ruthless power politics and did not regard international relations as an anarchic war of all against all. He saw international law, diplomacy and commerce as generally positive ways to keep the world as stable as possible. He regretted that war was inevitable but argued it should be limited by the principles of a just war.

Adam Smith (1723–90)

Adam Smith was a bit younger than Hume but the two were great friends throughout their adult lives. They admired each other's work and joined the same clubs and societies. Smith is often thought of as an economist, but he was a professor of moral philosophy and later became a private tutor to a young Henry Scott, the 3rd Duke of Buccleuch. This period is often referred to as his *Grand Tour* but was spent mostly in France. He ended his working life as Commissioner of Customs collecting import duties for the government. This is seen as ironic given his pleas and writings in favour of free trade.

Smith saw the world as a collection of different nations, or countries, all with different laws and customs, following local variations. Nations were an object of human passion and the honour of a nation was part of the honour of its people. Smith thought that people would always feel displaced in a foreign country, no matter how polite and human the local people were. He viewed the idea of world citizenship as largely erroneous, believing that ties with smaller social units were far more important. In other words, it was natural for humans to put family, friends and the nation first, and also in that order. He believed that love for your own country and love for humankind were two different things. While they both existed, he saw countries as being loved for their own sake, not as a part of a greater society of humankind. Smith was not sure such a society existed, but if it did, it would be best served by individuals who would direct their love toward a particular portion, i.e. their nation, since this was within their capabilities and understanding as well as in line with human affection (also chapter 6).

Smith believed that the idea of universal benevolence could not be stretched beyond one's country, but recognised that the idea of good-will knew no boundary, and could even include the entire humanity. A wise person should be willing to sacrifice her own interest for the public good of her country. Only God was 'the administrator and director' of the universe, who would take care of the 'happiness of all rational and sensible beings.' He confined man to 'the care of his own happiness, of that of his family, his friends [and] his country' (Smith 1982).

Smith did not see international affairs as a Hobbesian perpetual war of all against all. He believed that international order was possible and needed to secure stability and economic openness, which was the best way to serve individual liberty. He had less faith than Hume in international law. Smith thought that it was good to have international rules and regulations, but noted that international law did not secure – what was for him – the most obvious rule of international justice that only the warring party would receive punishment. He saw that being in full compliance with the international law of nations led to innocent people suffering instead of the leaders of states, who he saw as the guilty ones. During Smith's time *The Rights of War and Peace* (Grotius (1583–1645) 2005) was the most complete work on international law and even today remains an important text in the history of international law. Smith valued Grotius's distinction between rules on the basis of natural law and those based on positive law. He argued that the latter were inferior since the rules of international law were hardly ever reached by consent of all countries, let alone observed by all. The laws of nations were 'often little more than mere pretension and profession' (Smith 1982).

Despite Smith's doubts about the binding force of international law, he argued that sovereigns had a duty to attempt to maintain the common practices of an international society. He saw diplomacy as a way to promote continuing trade and to smooth international communications, and believed that immunity of diplomats and their residences should be a sacred principle. In return, diplomats should try not to offend their hosts. Diplomatic channels were important as levels of international commerce increased in order to reduce potential trade frictions.

Smith also viewed the balance of power as indispensable for international order, but wrote about it less than Hume (Van de Haar 2013a).

Smith, like Hume, regarded war as a normal feature of international life, since the principle of human sympathy could not be stretched indefinitely. In fact, Smith (1981) wrote quite a bit about defence, military organisation and the related issue of patriotic and military virtues in *The Wealth of Nations*. In a topical debate of that era about militias versus standing armies, he favoured the latter, in opposition to many friends and other public intellectuals. He believed in the principles of the division of labour and specialisation, arguing that militias, which consisted of part-time amateur soldiers with irregular training, would be inferior to standing armies. The defence of the country needed well-trained specialists, especially in a time when firearms and other weapons were becoming more technically advanced.

In line with the just war principle, he believed that war should be limited and its occurrence should be morally justified, defined as 'the abstention from injurious behaviour towards others' (Smith 1982). He believed that war needed to be justified as in a domestic court case and largely agreed with the Grotian rules for a just war including avoiding violation of property rights by another state, the killing of one's citizens, the imprisonment of one's citizens without recourse to justice, violation of one's territory, a continuous refusal to pay debts, violations of other contracts, conspiracies or the threat of violence towards one's territories. He also agreed that actions in war should also be justified and that civilians should be protected from retaliatory actions. Since wars were expensive, Smith proposed that they should not be paid for by incurring debt but by taxes directly felt by the public. This would certainly limit their scope and duration. At the same time, Smith did not think everything about war was necessarily bad. While not endorsing war, Smith felt that it should be recognised that war was also an opportunity for personal character building for a person of spirit and ambition. He believed that war enabled humans to learn to overcome the fear of death and led to military men developing the important virtue of self-command.

Smith is remembered for arguing in favour of free trade and against mercantilism, pointing out that free market and free trade offered the best opportunities for all people to improve their conditions. He saw government controls and 'beggaring thy neighbour' politics as plainly wrong and counterproductive and explained that free trade was not a zero-sum game. In general, even the most foolish ambitions of kings and princes could not match the immense welfare loss resulting from the jealousy of merchants and manufacturers. In every country it was in the interest of the majority of the people to have free access to free commerce and trade.

However, Smith also acknowledged the potentially negative effects of trade on international stability. As nation states became wealthier through trade, they would be more able to afford military equipment, support armies and engage in foreign wars. Without international order, there would be less scope for individual liberty to flourish.

He summed this up as 'defence is of more importance than opulence' (Smith 1981). He hoped that the combined efforts of diplomats and the military would ensure smooth foreign trade, assisted by the fact that trade also promoted cultural exchange, but still saw the commercial age as one with the occasional war.

Smith applied his 'system of natural liberty' internationally and was generally hostile towards imperialism, colonialism and slavery, both in the case of the American colonies and in the rest of the world. He argued that European empires were mainly founded on injustice and folly. Exclusive trading companies such as the British East India Company and the Dutch United East Indies Company were terrible monopolists and exploited and abused the local people. Their command was based on military force, and corruption flourished, not least by allowing humbly paid civil servants to engage in their own private transactions. He believed that the English were just a fraction better or less terrible imperialists since the few rules they could and did enforce were directed towards fostering natural liberty. Smith also saw empires as expensive to maintain and colonies as a disadvantage to all people concerned. For these reasons, he was in favour of American independence, writing that 'the rulers of Great Britain have for more than a century amused the people with the imagination that they possessed a great empire on the west side of the Atlantic. This empire however, has hitherto existed in imagination only' (Smith 1981). He entertained the idea of a federal union between America and Britain but realistically noted that this would go against British national pride.

To sum up, Smith held clearly developed views on international issues. He saw international affairs as based on human action, with all its positive and negative attributes. His international view was comparable to Hume's, favouring an international society of states to bring about international order and natural liberty, despite the occasional occurrence of war. He saw free trade as important but believed that it did not inherently lead to peace and he generally opposed empires and slavery.

4 AUSTRIAN SCHOOL: LUDWIG VON MISES AND F. A. HAYEK

In the history of liberal thought, the second half of the nineteenth century and the pre–World War I twentieth century saw the decline of classical liberalism and the ascendency of social liberalism in a number of countries. One notable exception was in Vienna in the late 1880s, when Carl Menger established what later became known as the Austrian School of Economics. He was followed by Eugen von Böhm-Bawerk and Friedrich von Wieser. Of the third generation, Ludwig von Mises was perhaps the most important thinker. In turn, he had a great influence on the fourth generation, of which F. A. Hayek was the best known member. While focusing on economics, the Austrians were also concerned with broader issues. Mises saw economics as part of a larger science of human action, which he labelled praxeology (Mises 1996) and which addresses issues beyond the usual economic arguments about the allocation of scarce resources. This wider science calls for considerations about the rules for government and living together, including the international domain.

Ludwig von Mises (1881–1973)

In interwar Vienna, Mises was one of the most influential and famous economists. He worked at the Vienna Chamber of Commerce, but also held a non-salaried *Privatdozent* position at the University of Vienna. He was a prolific writer and taught a number of thinkers who would rise to influential positions. Like many of his pupils he fled from National Socialism and ended up in New York City, where he resumed his teaching. Mises wrote two books on international relations, *Nation, State and Economy* and *Omnipotent Government* (Mises 1983, 1985), and covered international affairs in many other papers.

Nations were central to Mises's thoughts about international relations. Given his background in the scattered Austrian empire, it is understandable that he regarded them mostly as speech communities rather than strictly location-bound or blood-tied concepts. For Mises, the essence of nationality was language, although he acknowledged other characteristics such as geography. He saw a common language as binding the nation but rejected the idea of 'national character.' He believed that nationalism could be liberal and pacifist where the boundaries of the nation and the state coincided, such as in France and Britain. In these cases, nationalism was not directed against other countries. However, he also warned against a nationalism that could also be militant and imperialistic, which he believed occurred in areas of mixed populations, such as Eastern Europe. Mises thought the ugly side of nationalism should be solved by far-reaching self-determination. Even groups as small as several hundred thousand people should be allowed to form a sovereign state. These ideas obviously attracted criticism, including the charge that a Misesian system would lead to anarchy since the numerous states would constantly meddle in each other's affairs, disagree over borders, be influenced by geopolitical issues and be subjected to religious conflicts.

Mises regularly wrote and spoke about war, explained partly by his own experiences as an officer in World War I. His writings sometimes gave the impression that he entertained pacifist ideas, but he became increasingly belligerent over the course of his life. For example, he despised countries that wanted to remain neutral in World War II, because he saw it as support for National Socialism.

Like the Scottish thinkers, Mises believed that war was a deeply human activity. While there was nothing inevitable about war between certain groups, classes, races or nations, he believed that it would not become obsolete since there were too many potential causes of war. The reasons for war were not always rational. Wars due to religious differences or for nationalistic reasons were often waged without any real prospect of conciliation. He did not accept the criticism that his proposals for easy secession based on self-determination increased the prospect of war. He also disagreed with pacifists, who argued that it was impossible for people to gain from war. Military equipment should be procured based on free market transactions and there was no need for a centrally organised economy during times of war. Mises argued that the armaments industry was not a cause of war but one that responded to governmental demand. He believed that nations do not turn bellicose because of the interests of the weapons industry. This was in marked contrast to some social and economic liberals in the US, who warned against what President Eisenhower would call the 'military-industrial complex.'

Despite his awareness of the multiple causes of war, Mises's solution was exclusively economic. This imbalance meant his ideas on international relations were somewhat limited. His focus was mainly on economic nationalism and protectionism, with the international division of labour as the solution. He hoped that people would realise that their best interests lay in unhampered international transactions, particularly in trade and finance. His plea for free trade was almost absolute, including the free movement of labour. He viewed human history as a struggle between peaceful free trade and militaristic imperialism, with the latter regrettably most often winning the political argument. Nevertheless, he kept the hope that his classical liberal recipe would one day be implemented, so he always remained an idealist in this respect.

Mises stood in the just war tradition, but his dislike of the label 'natural law' prevented him from openly embracing it. He maintained that a war for self-defence was just and noted with satisfaction that even the strongest rulers responded to the need for a just cause for war. He felt that the just war tradition helped to create a situation where war was the exception rather than the norm. In times of peace, Mises supported international law that overruled national law to create a level international playing field in economics, although he also realised that international law was one of the least mature forms of public law.

Initially, Mises thought the balance of power in politics was a concept of the past. His view changed during World War II when he proposed the formation of two competing blocks in European politics, a Western bloc and an Eastern Democratic Union (Mises 2000; Van de Haar 2022). He argued that this would be the only reasonable way to reconcile the reconstruction of Europe, the world and the defence of civilisation. Given that he wrote this in 1941, he appeared to have anticipated a divide in Europe during the Cold War. He urged the US to leave its pre-war isolationist position and lead the way in establishing a free and peaceful world order, which was also in its own interest. He pointed out that whatever happened in the rest of the world would also be of great concern to America.

Mises was critical of the League of Nations, partly based on his own experience as a member of one of its committees. In 1919, he had already predicted its failure since it was built upon force, was badly organised and lacked an ideological foundation in liberalism. And the US was not a member, of course. He also recognised the League's useful activities, such as combating contagious disease, the illegal drug trade and prostitution, acting as an international bureau of statistics and developing work in the area of international intellectual property rights. As for the United Nations and other international organisations, he wrote that 'the spirit of conquest cannot be smothered by the red tape of international organisations, treaties and covenants' (Mises 1996). A world government would lead to world socialism, although he was open to the idea of a commonwealth of nations, as long as it did not become a global central planner or a unitary government.

Although he initially disliked plans for a (pan-Atlantic) Western Federal Union of democracies proposed by Clarence Streit, he supported these and other plans at the height of World War II. Mises became more positive about plans for Western and Eastern European integration, especially in the 1940s. He even served on a committee of the Pan-European movement after he arrived in the US. Consequently, in *Omnipotent Government* he argued in favour of a union of Western democracies to prevent Europe from slipping back again, as long as it was based on sound economics, knew no economic nationalism, had no trade barriers and had no bureaucracy. For these reasons, he opposed the Marshall Plan since he saw it as an example of misdirecting American taxpayers' money into all kinds of interventionist schemes. He devised his own plan for the establishment of an Eastern Democratic Union of Eastern European nations, which would actually be one big unitary state. This appeared to be inconsistent with his earlier argument for secession, most likely as a result of yet another devastating world war, and his strong conviction that the pre-war situation in Eastern Europe had been a failure due to the many minorities and nations that did not succeed in living peacefully, let alone according to sound classical liberal principles (Van de Haar 2022).

Not surprisingly, Mises strongly opposed imperialism, as he thought it was in essence a collectivist policy, treating people in the colonies as means, not ends. He also opposed development aid for former colonies that had become newly independent states. He preferred that they focus on building key social institutions such as private property, economic freedom, capitalism and an entrepreneurial spirit instead of becoming dependent on taxpayer-funded government-to-government aid.

From these few insights and references, it was clear that Ludwig von Mises gave serious consideration to the application of classical liberal principles to the international arena. His thoughts about the nation and secession developed over time, and he became less dovish. For over two decades he thought that supranational cooperation and federation could be a solution for unstable regions, as long as it was done on the basis of sound economics and other classical liberal principles. After World War II, he never returned to this subject, focusing on global state-dominated politics, such as the folly of development aid.

F. A. Hayek (1899–1992)

F. A. Hayek was arguably the most important classical liberal of the twentieth century. Although a Nobel Prize–winning economist, his work was much broader in scope, including political and legal philosophy, politics and international relations. In his view, it was the task of economic theorists or political philosophers to also attempt to influence current affairs. His ideas on international relations can be found in his books and articles, but also in newspaper opinion pieces (also see Boettke 2019; Butler 1985; Caldwell and Klausinger 2022) .

Like Mises, Hayek served as a soldier in World War I in an Austrian regiment where eleven languages were spoken. It made him realise that nationalism was one of the main reasons for the collapse of the Austro-Hungarian empire. This experience also underlined for Hayek that 'the group' was important for the individual. In international politics, the most relevant group was the nation, which he thought of as 'a homogenous community'. The nation was an important part of an individual's identity. Nations were prime sources of human organisation and individual loyalty. He associated nations with a common culture, defined as the same style of expression, both verbal and non-verbal, clothing style, food and other traits. He saw identity as a strong motivational force in human behaviour. Hayek thought that people would not tolerate a long-standing domination by groups of a different nationality. Hayek did not have any problems with patriotism but wrote about 'the poison of nationalism' (Van de Haar 2022).

Hayek was by far the most hawkish thinker of the four classical liberals presented here. He did not regard war as a regular policy instrument but saw it as an inevitable feature of life, arising from human nature. His bottom line was that there could be no liberty without safety, therefore international order was of the utmost importance. He even went so far as to allow the temporary sacrifice of basic human liberties in times of war. However, like Mises, he saw no reason for wartime economic centralisation. He was against economic intervention even for a country at war. Later in life Hayek became more belligerent. In several co-editorials, he supported US President Reagan's plan for increasing the level of defence expenditure, arguing that world peace depended on the US staying strong. He endorsed the policy of nuclear deterrence, supported the boycott of the 1980 Moscow Olympics by Western nations, and called for American intervention in the Iran hostage crisis. Hayek regarded pacifism as one of the main causes of war and supported the British during the Falklands War, arguing that the Argentine government broke many long-standing rules of international law. In these opinion pieces, he repeated his central argument that individual liberty depends on the maintenance of international order, with a central role for state violence if needed. He was less concerned than many other liberals about the potential breaches of liberty this might entail.

Hayek's preference for federation was another persistent feature of his international thought (Van de Haar 2022). In the 1930s, he joined the debate on possible forms of European or Western federation and supported his LSE colleague Lionel Robbins, the Streit plan for Western federation and even Norman Angell's idea to federally unite France and Britain. Hayek even proposed a full monetary and economic union between the two countries because he thought this was the best way to win the forthcoming war (which seemed inevitable to many people in the 1930s). International pooling of military resources was a good idea in some circumstances and could even lead to the decentralisation of a state's tasks, since he saw national defence as a centralising force in domestic politics. Later in life, he proposed plans to several leaders in Israel and the Middle East for the federalisation of Jerusalem and/or parts of Jordan, the East Bank and Israel. Hayek felt that Jerusalem should become a city modelled after Washington, DC, with free access for the citizens of the constituent member states. The Israeli politicians he approached with these ideas were less enthusiastic.

The Hayekian logic behind this was to regard federalisation as a last resort, a special solution to a political conflict that has yet to be resolved. He saw federation as enabling peaceful cooperation because federation presumed central federal control over traditional sovereign tasks such as defence and foreign policy, as well as an economic union. The main task of the federal government was to reject measures that would entail possible harm to its constituent parts. This way, a relatively orderly situation could be created, not out of a constructivist preference, but inspired by the wish to achieve international order.

While Hayek was not always careful in his wording, causing occasional confusion about his intentions, careful analysis shows that he did not favour international organisations or international law that were directed towards a constructivist goal. In the 1940s, he rejected the United Nations on the grounds that it suffered from the same problems as the League of Nations in that it was too large, lacked power and aimed to bring together countries that were too dispersed geographically. He felt that expecting an international organisation to make war impossible in the world was folly, since international organisations might also become sources of international friction. In the 1970s, Hayek criticised the UN as ineffective and constructivist. Unsurprisingly, he also criticised the International

Labor Organization. He saw the UN as a failed attempt to mix the Western liberal democratic tradition with Marxism. He was not against all international organisations, but preferred those whose remit was limited, analogous to a limited state in the domestic situation.

While he saw the Universal Declaration of Human Rights as containing important classical human rights, he regretted that it also contained numerous social and economic rights that were ill-defined, absurd and unenforceable. Like many classical liberals, he did not see human rights as positive rights, i.e. rights to, and suggested that international law would only limit the powers of national governments to do harm 'if the highest common values are negatives, not only the highest common rules, but also the highest authority should essentially be limited to prohibitions.' He regretted the fact that concepts of sovereignty and state had become instruments of legal positivists.

Hayek exposed himself to criticism when he travelled to General Pinochet's Chile and apartheid South Africa in the 1980s. Although he never supported the politics of these regimes, he defended them by arguing that they were not as bad as socialist dictatorships such as those in China, Libya, Algeria or Uganda, which he felt were not criticised as much. While arguably this may have been true, it was not a strong defence. He was enthusiastic about the economic policies of the Chilean dictator General Pinochet, but as Margaret Thatcher remarked in a letter to him, Chile 'had indeed gone through some economic successes, but some of its other policies were entirely unacceptable' (see Van de Haar 2009).

Hayek favoured decolonisation on principled, political and economic grounds. Like Mises, he strongly rejected government-to-government development aid and was influenced by the pioneering work of Peter Bauer. He believed that bad economics would keep poor countries poor and that the best way for them to get richer was by adopting better economic and social policies. He believed that aid did not promote development or alleviate poverty, but allowed too much meddling by donor countries in the internal affairs of recipient countries. Hayek saw free-market capitalism as enabling development, and the protection of property rights and economic freedom would unleash an entrepreneurial spirit, leading to innovation and economic development. He did not believe in any right to international redistribution of wealth, seeing it as immoral and failing to achieve its desired effect. Hayek also opposed policies that argued for limiting population growth, arguing that capitalism indeed gives life. In this, he was influenced by cornucopian environmental economist Julian Simon, who believed that if resources become scarce as a result of population growth, this increased the incentives to find alternatives or better technology to solve the problem. Human will, genius and determination were the 'ultimate resources' (Simon 1996).

As a strong free trader, Hayek believed that all international economic barriers should be removed in order to achieve the highest material welfare for all people in the world. He supported the progress of several negotiating rounds of the General Agreement on Tariffs and Trade (GATT), which led to a fall in trade tariffs. He believed that economic nationalism would lead to war and warned against monetary nationalism, arguing in favour of the denationalisation of money (Hayek 1990).

While Hayek was in favour of open immigration in theory, he warned that it would not work in practice. He feared the destabilising effects of a large influx of immigrants in a short space of time when entering a relatively homogeneous territory. He felt that this would lead to nationalist reactions and ultimately xenophobia, which was worse. He believed that a policy of more gradual immigration was the best way to deal with this issue, as he wrote in a series of co-eds in *The Times* (Hayek 2022).

Conclusion

These summaries of the international thought of four great classical liberals necessarily only scratch the surface. Some of their fundamental ideas were touched upon, such as their views on human nature, the possibility of ending war, means to preserve international order, or the political effects of free trade. In the next part, these ideas will be elaborated upon in more detail, sometimes also with reference to the work of Hume, Smith, Mises and Hayek. It will lead to a presentation and analysis of the building blocks (Freeden's 'concepts') of a genuine classical liberal theory of international relations.

PART II: BUILDING BLOCKS

5 INDIVIDUALS: HUMAN NATURE, NATURAL AND HUMAN RIGHTS

International relations and foreign policy are forms of human relations (Jackson 2000). IR scholars discuss the global political and economic system, including but not limited to the role of states as influenced by institutions such as regimes, international law or the balance of power. Indeed, some of these ideas are covered in this book. Yet, it must not be forgotten that international relations are predominantly about people. It is human action that makes the world go round.

Liberalism is the political expression of individualism in both the domestic and international spheres. As indicated in the introduction, this prompts the questions: what is the classical liberal view on human nature and individual rights and how does this relate to international relations? In this chapter, an answer will be given by focusing on three related issues: human nature, the consequences of the classical liberal view on human nature for its view on international relations, and natural and human rights.

As a political theory of individualism, every liberal theory has to start with the individual, and this also applies to an international theory. An important question is how the individual is perceived, how human nature is judged to be. What are a person's capacities and her limits? There are two aspects to this: the physical and the mental side. The former is a given and – despite all the differences – will not often influence how an international theory is constructed. The latter is more relevant.

'The science of man'

Most political ideas depend on how humans are seen. What are their capabilities – are they resilient, agile, strong, weak, social, anti-social? What is the role of the emotions, of reason, etc.? Political debate is often a struggle between underlying views of human nature since various political theories and ideologies are based on differing views of human nature.

Human nature was a key issue during the Enlightenment, the period widely regarded as the origin of classical liberal thought. Many Enlightenment thinkers tried to answer the question: what is human nature? This led to the development of a 'science of man', as Hume put it, which was the basis for understanding the social world and was meant to be comparable to how natural scientists, such as Newton, had uncovered truths about the natural world. An important idea was the shared conviction that, despite all the differences between people, there is still great uniformity in people's actions around the world and since humankind's beginnings. Human nature is seen as stable, and thus the knowledge of human nature is important to explain people's behaviour.

David Hume and Adam Smith made important contributions to this endeavour. Hume wrote *A Treatise on Human Nature* (1738–40) and tried to further explain his ideas in *An Enquiry Concerning Human Understanding* (1748) and *An Enquiry Concerning the Principles of Morals* (1751). Adam Smith was most famous among his contemporaries for *The Theory of Moral Sentiments* (1759), a catalogue of all aspects of human nature, such as sympathy, the passions, the virtues, the sense of merit and demerit, justice and duty. The question of human nature has remained a focal point in classical liberal thought. Both Ludwig von Mises in *Human Action* (1949) and F. A. Hayek in *The Constitution of Liberty* (1960), for example, started their economic and political ideas with the analysis of the fundamental capacity of human reason and its limits.

Classical liberal view

What comprises the classical liberal view on human nature? Classical liberals have a realistic view of human nature. They see humans as a social beings who function in groups and are largely dependent on others for their survival. This is significant because critics falsely characterise classical liberalism as fundamentalist with regard to individualism. However, classical liberals do not view humans as egoistic or atomistic, only interested in themselves, or without consideration for others. Anyone who makes that criticism is being deliberately misleading or has not read the great works of classical liberalism.

Classical liberals do not idealise human nature, which means they do not work towards, or base their ideas on, some better version of human beings. Humans are taken as they are, and human behaviour can be explained by an interplay of emotions and reason (besides the obvious physical limits even the strongest person has, as noted above). Classical liberals think there are limits to human reason, for example, in the ability to process much information, the inability to avoid mistakes or the inability to control emotions all the time. Humans are seen as ingenious compared to (other) animals but are also fallible and unable to 'maximise their utility all the time'. In fact, different individuals will have different views on their welfare and utility. The human individual is flexible and able to adjust to changing circumstances. So, there is not just one path in life to walk or just one way to use one's talents and become happy, contrary to what John Stuart Mill argued. Indeed, there is no such thing as a 'predestined path of life' at all.

The capacity of the human mind is amazing, but also limited. Classical liberals think that even the cleverest people are unable to plan such complicated systems as economies, let alone societies (Hayek 1993). It is just not feasible, and they have argued against all political theories based on these ideas. Communism and socialism are prime examples, but social liberals are also far more optimistic about people's rational capacity and the ability to change society on a rational basis. Emotions are important factors in explaining human behaviour. Hume even contended that 'reason is and ought to be the slave of the passions.' Morality, or questions about right and wrong, are founded on taste and passion, not in rational design, he argued (Hume 2000).

Classical liberals believe that humans are unable to always do the right thing. Conflicts, fights, violence and immoral behaviour are fundamental human traits. This does not mean that humans argue, fight and misbehave all the time. Most people value social order to enable a fruitful and meaningful society where they can use their talents in ways they see fit. However, it is illusory to think that all humans will always be able to restrain their emotions for some higher, long-term goal, for example, world peace, or even that they will all be able or even willing to abide by the rules to keep a society together; hence the need for law enforcement and a judiciary in the domestic situation. This fundamental insight also has consequences for conduct in international relations.

Consequences for international relations

It is impossible for humans to eradicate conflict and violence. This is an unfortunate but simple fact. Historian Margaret MacMillan shows that war is the fundamental trait in the history of humankind (MacMillan 2020). War is not an aberration, nor is peace the normal state of affairs. Human society and war are deeply connected. If you want to understand the world, you need to know about war. Classical liberals see that as evidence for their view on human nature and do not expect violence to be eradicated in the future either. This does not mean that classical liberals are in favour of violence; it just means that they take a more realistic view of humankind. For them, the relevant question for IR is how to deal with this basic fact of human life, which is confirmed by some modern research on human nature. The classical liberal theory of international relations is what Kenneth Waltz called a 'first image' theory, or a theory that 'finds the locus of the important causes of war in the nature and behaviour of men' (Waltz 1959).

The history of the human race is a history of war, as Churchill remarked. Steven Pinker notes that this state of affairs has stood the test of time, whether it be interstate war, ethnic strife, feuds, turf battles, etc. (Pinker 2002). He argues against the widely acknowledged claim that 'violence has nothing to do with human nature but is a pathology inflicted by malign elements outside us.' For Pinker, violence is not learned behaviour; it is part of human design. One cannot understand violence without a thorough understanding of the human mind, although social and political problems also play a role.

In a later book, Steven Pinker acknowledged that the use of human violence in general (including domestic violence, both privately and publicly, and international violence) has declined if one takes a long look at human history (Pinker 2011). This decline in violence is due to a number of external factors and institutions that have encouraged people to behave less violently, including rules set by states. However, less violence does not mean no violence. By contrast, MacMillan points out that the evidence for Pinker's theory is still debatable. Even if he is right, it is not reassuring, because the number of deaths due to war is still staggering. MacMillan refers to a long-running project of the Swedish Uppsala University that estimates that between 1989 and 2017 over 2 million people died as a result of war, while since 1945 some 52 million people have been forced to flee because of violent conflict. In other words, war is still real and it is not confined to the developing world, as the current war in Ukraine makes very clear.

Approaching this question from another angle, Richard Wrangham, professor of biological anthropology, agrees that humans went through a process of human domestication over hundreds of thousands of years to enable them to live together. There has been a decline in human violence as well as in the proportion of deaths. Yet that is only one side of the coin. Humans are still a dangerous species, in need of 'strong institutions, alert engagement to temper the rise of militaristic philosophies, the spread of excessively optimistic pacifism and the abuse of power' (Wrangham 2019). Humans have a low propensity for reactive aggression and a high propensity for proactive aggression. This means that long periods of peace between people do not have predictive power. Human capacity for organised violence may have diminished over time but that does not mean that it will disappear anytime soon.

Therefore, classical liberals prefer to concentrate on the foreseeable future. Authors with a specific focus on international affairs, such as Rosen (2005), Thayer (2004) or Rubin (2002), agree. They contend that biological and cognitive factors play an important role in predicting human behaviour, particularly that of political leaders. It cannot be ignored as 'politically incorrect' as has often been the case since the beginning of the twentieth century. As Thayer stresses, evolutionary theory provides evidence for human behaviour. It does not predict particular events but gives insights into human behaviour that help us understand the origins and causes of such events, not least warfare.

There is also modern research that relates to aspects that we may find outdated at first glance. IR scholar Michael Donelan wrote a fascinating book about the continuing role of honour in foreign policy, better known under modern alternatives, such as status, fame, admiration, glory, prestige or respect (Donelan 2007), while Christopher Coker pointed out the importance of culture and evolution in war (Coker 2014). All this confirms the classical liberal idea that humans have to deal with the inevitable outbursts of war and violence in international relations.

Natural and human rights

There are also other ways that 'the international' and 'the individual' cross paths. Classical liberals believe in the importance of individual rights and view these rights as being of a negative nature, so they argue that the state and individuals should not interfere with individuals' rights to property, freedom of expression, religion, freedom of association, etc. The use of the term 'negative rights' is not a value judgement of these rights but simply suggests what the state or individuals should not do.

In the nineteenth and twentieth centuries, other rights were added, known as positive rights, since in contrast to negative rights, which restrained the state or individuals, these 'positive rights' called upon the state to take action. Examples included the right to employment, to social security, to healthcare, or to regular holidays. Within liberalism, most of these rights are associated with social liberalism. They often require the infringement of classical human rights. For example, when a government demands taxes or collective premiums to finance the execution of these rights, it violates the right to property. This does not mean that classical liberals oppose all taxes. In fact, classical liberals differ among themselves on what the state should provide and whether that should be at the local, national or supranational level. However, most agree that the state should provide law enforcement, a judiciary and national defence, as well as a limited number of other tasks, for example in (basic) education, environmental protection and protection against uninsurable risks in the fields of health care and social welfare.

Rights are never absolute; one person's freedom ends where another person's freedom begins. In 1859, John Stuart Mill called this the 'harm principle' (Mill 1989). To give a famous example, freedom of opinion does not allow you to cause panic and havoc in a sold-out theatre by shouting: 'fire, fire'.

How does this discussion relate to international relations? Isn't the protection of human rights largely a matter of domestic politics? That is true, but there is also a large international component. Foremost, human rights are written down in international treaties, such as the Universal Declaration of Human Rights (1948) or the more extensive European Convention on Human Rights (1950) by the Council of Europe. Also, international human rights courts and tribunals have been established, such as the temporary International Criminal Tribunal for the former Yugoslavia, or the permanent International Criminal Court.

There are also questions of foreign protection (or enforcement) of human rights on the international agenda. Should the maltreatment of citizens of a country by their government be a concern for the international community? Should states intervene on behalf of citizens whose rights are abused? How about military intervention in case dictators kill their own people or commit genocide or related abuses? The United Nations refers to this as the Responsibility to Protect principle, or R2P (United Nations 2021).

Another way human rights play a role in international relations is with regard to the question of immigration and open borders. Should there be a right for every person on the globe to move and settle freely anywhere they like? The classical liberal perspective on all these questions will be dealt with in subsequent chapters.

6 GROUPS: NATIONS, STATES, SOVEREIGNTY AND IMMIGRATION

Classical liberals regard humans as social beings who live in groups. So, what role do groups play in the international arena? This is an important question for an understanding of (classical liberal) international relations. In this chapter, the most important groups in world politics will be introduced, as well as a number of ways in which states can be organised internationally. In addition, some important related concepts will be introduced: nations, states and countries; empires and federations; the concepts of national pride and nationalism; the role of diplomacy; and the idea of sovereignty. Lastly, the issue of immigration is discussed, or the desire to leave one group for another.

International politics usually involves a consideration of 'us' and 'them', i.e. the relations between different groups of individuals. Individuals involved in international relations almost always represent groups, especially states. The relationship between states and classical liberalism is not often discussed, certainly not in an IR context. One exception is Conway (2004), who argues that there is a compatibility between the equal moral standing of all human beings and their enjoying particularistic nationalistic attachments and affiliations. Some readers will be surprised to learn that classical liberals do not necessarily see the individual as the primary actor in world politics but rather the (nation) state acting on behalf of individuals and their liberty.

The international arena is a mix of governmental and non-governmental international actors, organisations and institutions, and domestic actors and institutions, as well as international and domestic rules, regimes and laws. National representatives are often intermediaries between the two domains, but there are also international bureaucrats in international bureaucracies. In general, national states still hold the majority of the power.

Nation, state, country

It cannot be emphasised enough, since it is a cause for so much misunderstanding: the classical liberal view on human nature regards humans as social beings, not as lonely hermits thinking about and caring only for themselves, as critics claim. This means that, despite their 'individuality', individuals cooperate with others to form groups.

Generally, the prime group for any individual is their core family, which is the natural group one is born into. It provides protection in the most vulnerable years or situations for most people. Broken families and people suffering from abuse and other social problems are, fortunately, still a minority. From there, individuals form bonds, often weaker but still significant, with their extended families, friends, neighbourhoods, villages, towns, regions or provinces, and the nation or country.

Hume noted that nations raise passions in all people, including 'the passion of national pride'. He observed that there are hardly any people to whom their nation is indifferent, and in 1751 even wrote 'I will joyfully spill a drop of ink or blood, in the cause of my country' (Hume 1932). Feelings for the nation are strong motivational forces for human conduct, both negative and positive. Yet he warned in his essay *Idea of a Perfect Commonwealth*: 'all plans of government, which suppose great transformation in the manners of mankind are plainly imaginary' (Hume 1985).

Smith largely agreed and underlined that nations are kept together on the basis of sympathy. Yet people have the duty to put relatives, friends and country first, in that particular order. In his mind, there was a difference between love for your own country and love for mankind. The former was natural and feasible; the latter was impossible for mere humans, and therefore should be left to God. The nation or country, as a cultural unit defined by language and/or cultural descendancy, was the largest group that provoked feelings of sympathy and adherence.

One can feel sorry for people in other parts of the world but these feelings are often temporary. Smith gave the example of an earthquake in China: initially, people would be shocked to learn about such a natural disaster, but they would also very quickly move on with their own lives. They are far more concerned with relatively minor events occurring in their vicinity. This old insight is confirmed in our time, for example, by the way people react to natural disasters around the globe. Initially, we are shocked and might donate some money to an organisation offering relief, but after a few days, the disaster is no longer talked about, even though the effects remain very real for the people concerned.

As we have seen, Mises counted feelings for the nation among the strongest feelings of any individual and wrote two books on the nation and related issues, namely *Nation, State and Economy* (1919) and *Omnipotent Government* (1944). He personally felt the negative consequences of nationalism throughout his life, yet never endorsed the idea of a 'nationless world'. His preferred solution lay in the secession of groups large enough to form an independent state themselves, thus arguing for a world of even more states. Hayek was less convinced about this, but saw the emotional attachment to a nation as an important element of individual identity. For Hayek, the nation state was the relevant group in international politics. He thought this was fundamental because people would not tolerate long-standing domination by groups of people of a different nationality. Like Smith, Hayek argued that humans had weaker emotional bonds and lacked deep feelings of responsibility for people living further away. He acknowledged that it would sometimes be impossible for nations to live together in relative peace and order. In such instances, a (interstate) federation might be a solution (also see Van de Haar 2022).

A nation does not always equal a state. The distinctions are fuzzy, but in general, a state is a political unit while a nation is a cultural phenomenon. The border of a state hardly ever fits perfectly with the nation. Often there are more nations living within a state or there are cross-border issues. The position of the Kurds, a nation without a state, comes to mind. The old classical liberals mostly wrote with 'old Western Europe' in mind, although they included nations such as Japan, China or Siam (Thailand). But countries such as Germany and Italy were only united in the late nineteenth century. Every state has its own history. Statehood is fostered by telling myths, through formal education, the introduction of national symbols and the concept of a shared national history. It should also be recognised that our current understanding of nations and states is based on a Eurocentric Westphalian view. In other parts of the world, such as Africa and the Middle East, states were created by colonial powers by drawing almost random lines on maps, often ignoring ethnocentric loyalties.

There is no denying that in international affairs groups are important actors. Classical liberals acknowledge and support that, as long as these groups act on behalf of, or fight for individual liberty. In the eyes of many classical liberals, there is a natural relationship between human nature, the nation as a cultural community and states. The world is dominated by states and they are the prime actors in international relations. Classical liberals take this as a fact and regard states as the natural political units at the international level because at heart this usually follows from individual emotional attachments. Many people share the Smithian 'love of your own country' in preference to 'a love of mankind'.

National pride and nationalism

Countries or states are political units, but they also raise feelings of passion in most people. On the other hand, negative feelings may occur when the nation is humiliated, for example, when a war is lost. Smith thought that most people feel displaced in a foreign country, no matter how polite and friendly the local people are. Hayek thought that people do not tolerate any long-term dominance by groups of a different nationality. While these generalisations were made some time ago, the emotions they describe are still very recognisable. Hume's conclusion that 'there are few men to whom their country is in any period entirely indifferent' still stands (Hume 2000).

These feelings for the nation were – in the eyes of classical liberals – abused from the late nineteenth century onwards, when they became a cornerstone for the rise of nationalism. Chauvinism was just a harmless overestimation of one's nation's qualities and achievements. However, nationalism is a collectivist idea. It places the group's well-being above individual liberty. Therefore, it can never be part of a classical liberal theory of international relations.

Sovereignty and diplomacy

How do states relate to each other? Here we touch upon the discussion of sovereignty, which Jackson (2007) defines as:

> an idea of authority embodied in those bordered territorial *organizations* we refer to as states, and is expressed in

their various relations and activities, both domestic and foreign. It originates from the controversies and wars, religious and political of sixteenth and seventeenth century Europe. It has become the fundamental idea of authority of the modern era, arguably the most fundamental.

Sovereignty is an idea that refers to both the supreme authority of the state and the political and legal independence of geographically separated states. It is an idea concerning the rights and duties of governments and citizens or subjects within particular states. Yet it is also an idea about international relations. States stand in relation to each other, each one occupying its own territory. They deal with each other through diplomacy and other forms of foreign relations.

Many classical liberals are concerned about sovereignty, as it is, foremost, a means of protection of life, liberty and property. Within a system of states, the idea of sovereignty is a regulatory measure; states should respect each other's sovereignty, thereby preventing international anarchy. It is an important institution in international relations that protects states and their inhabitants against each other. It also helps secure the rights of smaller states against larger or more powerful states.

Of course, sovereignty is not always a positive concept. There are many examples of rulers who mismanage their countries and violate the freedom of their citizens. Such rulers tell foreign governments that they should not interfere in any way since that would be an infringement of their sovereignty. Classical liberals acknowledge this downside but think that, on balance, the advantages in the form of a stable international order are greater, especially for fostering individual freedom, than the regrettable disadvantages to the inhabitants of that particular country. Of course, countries always run the risk of foreign military intervention if they oppress or kill their citizens. This will be dealt with in the next chapter.

Diplomats are a key channel for maintaining relations between sovereign states. Smith and Hume valued the role of ambassadors – resident or travelling – as channels of communication between country leaders. Special rights and immunities should be granted to smooth their operations. As Watson (1982) pointed out, diplomacy is much more than resident embassies and professional diplomatic services. Nowadays, there are more channels and actors involved in international dialogue, such as non-governmental organisations, yet diplomats have retained their importance and relevance. While our modern world is much different, especially in terms of transportation, the media, the meetings of leaders and other means of communication, the role of diplomats has at its core remained stable. Diplomats open doors that remain closed to others and attempt to smooth relations between countries. With their access to the highest government circles, they are able to prevent miscommunication while taking care of national interests along the way.

Empire

Empires have long been dominant across the globe; for example, China, Mongolia, or the Latin American empires before they were conquered by the Spanish and Portuguese empires. Historically, the European empires have been the most dominant and their traces can still be found, such as in the French, English and Dutch islands in the Caribbean. In the nineteenth-century 'scramble for Africa', the Germans, Italians and Belgians also played their parts, while the Russian empire was landlocked.

As we have seen, on the then-contemporary issue of American independence, Hume and Smith supported the Americans. Both believed it was not good to dominate other peoples. They also did not favour the empires European powers were building in other parts of the world but were not vocal opponents. Smith did argue against the monopolist trading companies, such as the Dutch East Indies Company, that in effect ruled whole parts of the European empires and were treating local inhabitants badly. They only focused on their trade and hardly cared about the people. In the twentieth century, during the post-war decolonisation, Hayek and Mises saw a clear relationship between nationalism and empire. Those who wanted to keep the empire often argued in terms of 'keeping the mother country great'. This is a collectivist line of thought, which classical liberals refute. The bonds of former colonies with the mother nation do not depend on keeping an empire. Ruling over other nations and their citizens and taking away their (natural) resources is bad. It is unethical and continues to be opposed by classical liberals.

Also, classical liberals do not think there are 'supreme' countries that have a right to rule others, just as they do not think there are natural leaders who have a right to dominate others in a domestic situation. Therefore, only voluntary agreements and treaties should form the proper basis of dealings between countries in international relations. On this basis, Deepak Lal, a classical liberal economist, would be seen as being far too positive in his *In Praise of Empires* (Lal 2004) when he pointed to what he saw as the positive role that empires played in providing the order required for peace and prosperity. This may have been true for some empires at some points in history, and it is easy to overlook that in some colonies where Western empires introduced good education, healthcare and infrastructure, at least for some, but equally in others there are stories of oppression and the undermining of rich indigenous cultures. Arbitrary rule over others can never be part of classical liberal international thought.

Federation

While the state is the primary actor in world politics for many classical liberals, there are situations where sovereign states simply do not get along in a peaceful way and wage war far too often. In such cases, classical liberals think it is better to form a federation, where nation states pool some of their powers, either economic, military, political or combinations of these three. The idea is not to create a world government or to abolish national states, but rather to introduce interstate federation as a means to international stability. There are many forms of federation, but, for example, in the case of France and Germany, it was clear that after three wars in 80 years (1870, 1914,

1940) something had to be done. For them, the European Communities were a step in that direction.

As was previously mentioned, both before and after World War II, Hayek and Mises wrote extensively on this issue (Van de Haar 2022). Mises developed all kinds of federal schemes for Eastern Europe, was an active member of the pan-European movement for some time and supported federal schemes for Western Europe as well, as long as these federations were based on sound economics and did not have trade barriers or a large federal government. Hayek participated in this debate and gave much thought to the usefulness of federations throughout his life, including in the Middle East. In England, the debate about federation had roots in late-nineteenth-century proposals for the federalisation of the British Empire. It is important to note that in the classical liberal view, federations may be a suboptimal yet last-resort solution where nationalism, religion or other factors pose a threat to the international system of nation states.

It should also be acknowledged that this classical liberal federal solution to international disorder is not easily put into practice. The bottom-up pooling of sovereignty proves difficult in practice and classical liberals do not always agree with each other. For some the European Union is a good example as it evolved into a unique blend of federation and cooperation between sovereign states in favour of market liberalisation. For other classical liberals, the EU has become a bureaucratic protectionist top-down illiberal project. This was evident during the UK's referendum to leave the EU. There were classical liberals who supported remaining in the EU and others who supported leaving the EU.

The development of ASEAN in South-East Asia shows how difficult it is for regional organisations to grow into bodies that can overcome recurring military conflicts. The bottom-up approach, where sovereignty is strictly protected, forms the basis of most supranational regional organisations and does not necessarily lead to federalisation. Some but not all classical liberals would view a top-down approach as the least worst option after the end of a conflict, as part of conflict resolution or a peace plan. This underlines that federalisation is (most often) the exception to the classical liberal rule of a world of sovereign states.

Immigration

Since classical liberals defend negative human rights, there is some debate over whether freedom of movement should include the freedom to move freely across borders. Currently, under international law, states have the right to stop or limit immigration, but is this justified? The answer for classical liberals can largely be seen as a trade-off between the freedom of movement and the right to property. This is often a topic of debate among classical liberals, but also between classical liberals and libertarians (see chapter 10).

Although the debate over free immigration and open borders is topical, there is nothing new about it. Mises (1927) laid out the main contours of this debate from a liberal angle, arguing that, for a liberal, every person has the right to live where she wants since this is part of the essence of a society based on private ownership of the means of production. He felt that attempts to justify immigration barriers on economic grounds were doomed from the outset. There can be no doubt that barriers to migration diminish the productivity of human labour. Yet Mises recognised that the debate is not about economics alone. People living in states that are attractive to potential immigrants feel threatened by the idea of a large influx of foreigners. Immigration may, in some cases, make indigenous people feel that they become a minority in their own land. Mises gave the example of Australia's opening up to immigration from neighbouring Asian countries. He also felt that the US was only able to cope with the immigrants from Europe in the late 1800s and early 1900s due to this group's being small enough to assimilate into American culture. He also noted that the US subsequently introduced severe immigration laws. In a world where some states are powerful, it is understandable that a national minority may expect the worst from a majority of a different nationality. While Mises took a eurocentric view, it could be argued that the impact of the influx of European colonialists into Australia, the Americas and other countries on indigenous people certainly seems to justify this fear. Mises felt that the only solution to this fear of immigrant majorities would be the introduction of a completely liberal state, i.e. one with limited responsibilities and powers.

Yet as long as such limited states do not exist, pleas for free immigration are seen by some classical liberals as potentially posing a threat to individual liberty – not only to the liberty of the potential immigrants but also to the liberty of the incumbent populations. This is constantly overlooked or dismissed by liberal writers such as Van der Vossen and Brennan (2018), Somin (2020), Lomasky and Tesón (2015) and Kukathas (2021). Some of these authors are considered to be classical liberals. While their precise arguments differ, they all agree that it is preferable to allow individuals from poor countries to freely migrate to richer countries so that they can choose where to live and where to use their talents to lead the best life they can. They believe that this would be morally right and point out the positive economic effects in the short term for the migrants themselves, as they immediately improve their position, even if they are among the least well-off in the receiving country. In the medium to long term, the receiving country benefits too, as immigrants start to contribute to economic growth. They also believe there are possible schemes to compensate for other costs involved with free immigration, such as social welfare and health.

Besides the counterarguments posed by Mises, there is also a different issue, which the Austrian seemed to have overlooked: a classical liberal limited state entails a limited number of infringements of natural rights. In other words, these rights are not absolute. Therefore, limiting the right to immigrate is not necessarily illiberal. Classical liberal philosopher David Conway carefully debunks many of the arguments put forward in favour of free immigration (Conway 2004). He concludes that there is no sound liberal principle that forces states to treat citizens and immigrants with equal respect and concern. He denies that foreigners have a moral right to enter the public spaces within the jurisdiction of a foreign state without the latter's express or tacit consent. These public places may be said to be the joint property of its citizens for whom the state acts as a proxy. Immigrants have no more right to enter a foreign country at will than its citizens have to enter the private property of fellow citizens without their consent.

The fact is that no classical liberal state exists anywhere in the world. Milton Friedman (1978) made the point that where a welfare state exists, even if it is just a limited welfare state, free immigration becomes a burden to these arrangements. It makes it impossible to keep the welfare state in existence. Current writers attempt to counter this argument with compensation schemes, but they tend to overlook that immigration can still be regarded as a violation of the property rights of the receiving population, which is obliged to pay premiums for those services and build up certain rights accordingly. Unlimited immigration is seen to violate these property rights to welfare arrangements. Compensation is possible in theory, yet in practice they will still be paying for it, either directly through higher obligatory premiums or by enjoying less access as more people make use of welfare services. The chances of adequate compensation are not seen as great.

Generally, the position of the receiving population is not sufficiently considered in the debate on immigration. For example, Hayek wrote several op-eds arguing, on the basis of his own experiences in Vienna, that a large influx of immigrants would have an enormous impact on the culture and the lives of the receiving population. It might lead to protests and nationalistic reactions which, while not defendable, are to some extent understandable.

The above points should not be seen as an argument against all immigration. It is just a counterweight to the idea that a classical liberal should favour unlimited immigration on economic and moral grounds. While some classical liberals do, there are others who do not, based on consideration of property rights, culture and a minority–majority perspective. Many Western countries welcome a limited number of immigrants, depending on whether there are labour shortages or an ageing population. There is a moral case for accepting refugees fleeing war and persecution. Beyond that, classical liberals disagree on the level of openness to immigration. For example, many European classical liberals welcome freedom of movement within the EU but tend to favour only limited immigration from non-EU countries. No doubt, the debate will remain topical in the foreseeable future.

7 VIOLENCE: BALANCE OF POWER, WAR, MILITARY INTERVENTION

The chapter on human nature made clear that individuals and groups sometimes get into conflict. Therefore, this chapter will deal with the questions of what, if any, is the role of violence in international affairs and how it should be addressed. The general classical liberal view on international affairs and the way states deal with each other, particularly in the balance of power and in war, will be discussed. Then the rather particular position of classical liberals about the cost of war will be introduced. The chapter closes with a debate on the pros and cons of military intervention.

Many will see these as *hard-core* international relations issues, although they are not necessarily hard-core issues for liberalism (to be discussed in part III). But what should other countries do if sovereignty is not respected, if violent conflict breaks out between states and if a ruler commits crimes against her own people?

View on world politics

Classical liberals' view of world politics is parallel to and similar to their view of human nature. The most important observation is that there is no overarching power or a judge who determines what needs to be done or what the main rules are in international affairs. Of course, there is international law and there are international courts and tribunals, but it is important to realise that none of the existing legal institutions have the power to ensure their judgements are executed. In essence, it is about states, their willingness to cooperate and their position towards other actors.

In a world devoid of a supreme authority, all states face a security dilemma (Booth and Wheeler 2008). This means that they cannot count on the existence of a stable and peaceful order, even if such an order would be best for general human well-being. There is always the threat of a state, or a group of states, taking advantage of the absence of a global government. The security dilemma is therefore existential. States need to take care of their own security, first and foremost militarily, if they want to survive. Leaders and elites can never be certain about the intentions of leaders of other states, even when they have no intention of harming any other state. In the same vein, weapons that are procured purely for self-defence can be seen as offensive by others. Perceptions matter a lot in a world of uncertainty.

In line with domestic politics, classical liberals aim for a middle position where states provide some kind of order while acknowledging that anarchical conditions or a lack of order will exist to some extent. As shall be seen in part III, classical liberals favour 'the anarchical society of states', as the Australian international relations theorist Hedley

Bull (1995) called it. In this situation both power-based relations between states and rudimentary societal features such as international law and international cooperation exist. It enables states to live together in a relatively structured way that also addresses their security dilemmas. It is a political society between states not a multi-actor international society where all actors are on equal footing.

Balance of power

The international world is characterised by a degree of anarchy, i.e. a lack of order to some extent. Since this can be a potential threat to individual freedom, classical liberals attempt to take measures that foster as much international order and cooperation as possible. They attempt to achieve a balance between two potential threats to the core values of classical liberalism. One is pure anarchy, defined here as the rule of the strongest, without meaningful or lasting protection of the rights to life, liberty and property. The other is attempts to unite humankind under one power, such as the proverbial Hobbesian Leviathan or a world state.

An important element is the balance of power. Bull (1995) referred to this as an 'institution' and it has been a favourite instrument of international order for many classical liberals. Hume wrote a well-known essay about it, Smith agreed with it and twentieth-century thinkers Mises and Hayek also held it in regard (Van de Haar 2009, 2011). The origins of the balance of power go back at least to ancient times. The idea is almost self-explanatory: states have an interest in preventing domination by one big power, since this power would be able to execute its belligerent plans without serious opposition. Therefore, other powers form an alliance to balance the power of the dominant state or alliance of states. They balance each other out, and the result prevents major wars, since the costs to both sides are seen to be too high.

Little (2007) points out that the balance can be *associational* when it rests on an idea held by a group of leaders of great powers, for example, after peace has been negotiated at the end of a war, such as the Concert of Europe after 1815. An *adversarial* balance develops when it is aimed at checking the existing or rising power of another state and its allies. The most notable example would be the Cold War that lasted from 1945 to 1990, between the US-backed 'free world' on the one hand and the Soviet Union–backed communist world on the other.

Throughout his career, Hayek wrote about spontaneous order, which emerges as an unintended consequence of individuals or groups of individuals pursuing their own personal goals. Spontaneous orders evolve and lead to equilibrium between their constituent parts. The balance of power is a powerful example of Hayekian spontaneous order in international affairs. Some states want to dominate others, either alone or in alliances, but in that pursuit they come across other states that have the same goal, which results in an equilibrium of relative order and stability (Van de Haar 2011).

Nevertheless, the balance is not perfect since it is often fragile and hardly ever stable. States may change alliances, rise or fall in power, the influence of great powers may fluctuate, and fewer wars may be fought between members of alliances or their proxies. During the Cold War, there were many 'hot wars' and conflicts in Asia, Africa and Latin America. But, in general, the balance of power helps prevent one power from dominating the other and contributes to international order. The current military aid of Western countries to help Ukraine fight against the Russian invasion is another example. Classical liberals see these as examples of spontaneous order at the international level.

War

War is one of the most devastating events in human experience and one of the most serious threats to individual liberty. There are innumerable costs, including lost lives, offences against human rights, material damage, environmental damage, loss of homes and economic costs, to name a few. It is as true today as it was hundreds of years ago. Classical liberals do not take this lightly, nor do they only theorise about war in an abstract way. However, they consider the occasional occurrence of war inevitable. It follows from human nature, sometimes from aiming to maintain the balance of power, as well as other causes, such as religion, geography, raw materials, or unpredictable despotism in a particular country. Therefore, the question is not 'How can we get rid of war forever?', as some other liberals pose, but instead 'How should we deal with, and possibly limit, the inevitable occurrence of war?'

This is not meant to be fatalistic. Like many other people, classical liberals prefer to avoid war as much as possible. They have never regarded it as a normal Machiavellian policy instrument but have sought ways to ensure there was a justifiable reason for going to war, if it had to occur, and to regulate the way wars were conducted. Prudence, reasonableness and caution are the main virtues mentioned by classical liberals in this respect. Not surprisingly, there is a strong relationship between the just war tradition in international law and classical liberal thought about war. For example, great international lawyers, such as Suarez, de Vattel and especially Grotius, were held in high regard by Hume and Smith.

The just war tradition demands that a war have a justified cause, though the list of justified causes in the just war tradition is extensive. The main ones include the right to defend oneself in case of an invasion or other breach of the state's sovereignty or territory, the killing of its citizens by another state, when citizens are taken hostage, when debts are not paid, or the violation of (peace) treaties. The second criterion of the just war tradition is that wars should be conducted in a just way. A major rule is that innocent non-combatants (civilians) should be kept out of the hostilities. Also, when a certain territory is conquered, the leaders may be replaced, but the population should be allowed to continue their lives, religion and habits as much as possible.

Rengger (2013) emphasised that it is illusory to think that the just war tradition can solve conflicts or eliminate war. The tradition may be seen as a restriction on war, which is true compared to unlimited war waged by powers that are less or not concerned with a moral basis for war. However, the just war tradition has often been used as an excuse to go to war and for the way it is waged. Just war has developed into an idea which is more concerned with administering justice by a political authority rather than restricting the use of force. It is therefore no surprise that wars have increasingly been deemed as fighting for a just cause, especially after World War II. There has not been a similar adherence to the second principle of the just war tradition, which constrains the use of violence. Generally though, the just war tradition is one of restraint. Classical liberals recognise the importance of restraint but do not believe this means that they should be pacifists. Given classical liberals' views of human nature and of international affairs, pacifism is often seen as an unrealistic or even inappropriate response. War can sometimes be needed to avoid bigger problems in the future, for example to stop an authoritarian ruler from conquering more countries.

Most arguments about war in the wider liberal tradition usually consider a war between two or more states. However, there are many other types of war, such as guerrilla warfare or conflicts where proxies of states fight, or a combination of conflicts between armed groups, states and state-sponsored armed groups, such as in the recent wars in Syria and Iraq. In fact, there are probably more complex wars of the latter type than of the former type being waged in the modern world. In terms of academic study and practical politics, these differences are important. For classical liberals, these types of conflicts are mainly evidence of the inevitability of war, while dealing with them can be even more complex.

Cost of war

Within the classical liberal and libertarian traditions there is a sub-literature on the 'hidden costs of war'. These are rules and legislation, often illiberal, that are put into place by the state in response to the threats of war and to enhance the nation's chances of victory. Examples include the curtailing of free speech or free association. Most of the literature focuses on the economic implications of war, for example, when states nationalise whole parts of the economy as part as the war effort. Hayek and Mises strongly objected to such a war economy, arguing that in times of war, the free market continues to foster the most innovative and efficient producers. Higgs (2005) warned that these measures are never fully reversed after a war. Therefore, wars are incidents that lead to a larger state. As a consequence, Higgs calls for abolition of war. Classical liberals share the analysis but believe that the solution is unrealistic. War will not go away. Reclaiming the lost liberties after a war is important, yet mainly a matter of domestic politics. To call for the abolition of war because of this mechanism, as some do, may be sympathetic but also utterly unrealistic.

Military intervention

One of the questions that challenges classical liberals and others who believe in the just war tradition is the issue of military humanitarian intervention against leaders who persecute their people. This is usually framed as, should state A intervene in or invade state B to undo a serious wrong the leaders of state B have conducted against their own population, such as a grave infringement of human rights? After the end of the Cold War balance of power, such interventions have become more commonplace. The United Nations refers to this as the Responsibility to Protect principle, or R2P (United Nations 2021). R2P has been used to justify interventions against regimes who are considered to have committed human rights violations against their own people. Such interventions are no longer seen as affecting the global order, as was the case in the Cold War.

The question of whether a state or coalition of states should be allowed to intervene on behalf of a suffering population raises many issues including the ethical dimension. Since classical liberals believe in upholding individual human rights, shouldn't they also believe that there is a right, or perhaps even a duty, to intervene on behalf of foreign citizens, who are also humans with the same rights? To answer this question, we can explore the thoughts of Hume and Smith (Van de Haar 2013b). Both saw sympathy as a central mechanism for regulation and evaluation of human relations. The justification of military intervention is about sympathy and justice, and their application to human bonds. If bonds are stronger (sympathy), then intervention is perhaps more urgent and also possibly just (justice). Smith defined sympathy as 'people's natural inclination of fellow feeling with any passion whatsoever'. It is the capacity to put oneself in the place of another person. This capacity leads to a judgement (propriety) of certain behaviour, but also of feelings of compassion, such as love, sorrow, benevolence, pride, etc. Smith used this idea to introduce an imaginary 'impartial spectator', who would be the judge, or the yard stick, to measure one's own behaviour. Hume regarded sympathy as a 'principle of communication' which informed people about other people's inclinations and desires.

Hume emphasised a difference between specific and general forms of sympathy. The first was directed towards particular persons, the latter a more general feeling of shared humanity. This latter could not be a motive for action. As discussed before, there was no brotherhood of men and therefore no need, even if there was a possibility, to act in service of that alleged brotherhood. Only close bonds lead to particular sympathy, and feelings of sympathy become weaker if the bonds are weaker in terms of distance for the individual and their core groups. Sympathy did not lose all meaning across the border, but was certainly much weaker and hardly ever a motive for action. Temporary feelings for the misery of other people further away were of course possible but would, for most people, remain of an incidental nature. Hence, there was no moral duty for military intervention.

While there was no duty, both Hume and Smith were less clear about the right to intervene. They mostly regarded justice as a negative rule, about restraint and not doing things. Hume added that in international affairs the natural obligation towards rules of justice were less stringent than in the domestic situation. On the other hand, they also supported Grotian international law and thought international rules were useful to achieve international order. Sovereignty played an important role, and non-interventionism was the rule.

Still, rulers had to respect the rules of justice in international relations, even if these rules were considered weaker than for a domestic situation. Grotius allowed, yet never required, 'corrective justice', in the case of a tyrant committing atrocities against their subjects. However, this was meant to punish the ruler rather than protect the people concerned. While suggesting some justification for intervention, it did not constitute a right to intervene, especially when considered in combination with the demand for virtuous and prudent leaders, whose task was to maintain social order, both nationally and internationally. To sum up, there is also no strong classical liberal principle for intervention, although there is some room for it when a ruler seriously misbehaves. As Smith and Hume acknowledged and supported, this leaves some room for manoeuvre for the leaders of states.

These principles are not much different from the current situation laid down in international law. Sovereignty is still a major principle for the current world order, which mostly sees military or humanitarian interventions as unjust. A difference between our time and that of the Scottish Enlightenment thinkers is that the United Nations' commitment to Responsibility to Protect (R2P) does allow military intervention in certain circumstances. However, intervention in the internal affairs of a sovereign state is still the exception. Most classical liberals are comfortable with that idea, and, contrary to other liberals, do not necessarily promote more military or humanitarian intervention.

Effects of military intervention

Even if you are in favour of military intervention, does it really work? The military part of the intervention might well work, in the sense that the 'bad' regime is toppled, but politicians, military strategists and others have to consider both what happens next and the exit strategy, i.e. when to leave and how? The experiences of the last twenty years in Iraq, Afghanistan, Libya or Mali demonstrate intervention followed by attempting to build liberal democracies from the outside has largely failed. Despite valiant efforts, it has proved just too difficult to export a culture of liberal democracy off-the-shelf. Even less ambitious interventions have been led to UN peace-keeping forces remaining in a country for decades, as in Cyprus. There have also been cases where many people of different opinions would have agreed that intervention was necessary. The clearest example would be the 1994 genocide in Rwanda, when the major powers did not intervene and left it to an underarmed UN force on the ground. While the pressure to 'do something' will always be around, classical liberals often resist these calls for action.

Hume wrote that all government is based on opinion, or the idea that leaders ultimately depend on the consent of the governed, who are always able to topple any regime. That may take a long time in practice, in some countries even centuries, but history shows the possibility for revolution is real, everywhere around the globe. Friedman (1962) pointed out that individual liberty and economic liberty are necessarily related and that one cannot have one without the other in a 'good society'. It could be expected that in countries with some measure of economic freedom but without political freedom, the populace will eventually demand more rights, even though that may take a long time or may require a revolution. In the meantime, much misery may be caused.

This is not an argument for always doing nothing. It is just that some classical liberals see the proposed cure of military intervention as worse for all parties concerned or as leading to unintended consequences. The world can be an ugly place, and achieving some sort of stable international order is already a large feat in itself. The balance of power can help to prevent major wars and conflicts, but not always. Intervention is sometimes called for but does not change the fundamental principles underpinning global politics. Classical liberals attempt to deal with the reality that a harmonious world without conflict is an admirable goal but remains a fairy tale.

8 RULES: INTERNATIONAL LAW AND INTERNATIONAL ORGANISATION

Is there a place for international rules, and if so, what rules? Classical liberals do not think international relations is characterised only by violence. There is indeed a place for rules, even though international law is often categorised as 'soft law' while domestic law is seen as 'hard law.' The difference lies in the possibility of enforcing judicial decisions. Not all conflicts or issues lead to violence, most can be tackled by agreement between states, e.g. through a treaty or a set of international rules that apply to all parties concerned. Sometimes there is a role for international governmental organisations (IGOs), while non-governmental organisations (NGOs) also play their part. Classical liberals value these agreements as long as they help maintain an order that is needed for individual liberty. Promises, treaties and rules should be adhered to, by way of moral incentive. Yet, classical liberals also worry about the increasing number of international laws, rules and regulations. Sometimes they help to protect individual liberty but they can also endanger it.

International law

Treaties and trade agreements are probably the oldest forms of international law. They serve as expressions of common goals between countries, and set norms for the international society of sovereign states. They restore order, help to settle disputes and extend the rule of law to some extent to the international level. Nowadays, there are thousands of treaties and other forms of international law. States often comply with international law, judging this to be in their best interest, or a moral duty. In this way, international law helps establish and protect international order.

Yet international law often lacks an enforcing authority. States themselves enforce international law. Sometimes dispute resolution is part of a treaty or delegated to an international organisation such as the World Trade Organization (WTO). Or there is agreement as to where a dispute should be resolved, e.g. an existing court or arbitration body. Ultimately though, especially in conflict situations, even the highest international court can be seen as toothless. In those cases, other states can either give up, protest, put sanctions in place or ultimately go to war.

For classical liberals, good international law comprises rules to help states deal with one another, each other's citizens and resolve existing or potential disputes in a similar way to how such issues are resolved in domestic politics. It makes sense for states to cooperate on a number of international issues, such as travel, communications, border protection, trade, finance, tackling people trafficking and environmental issues, including the protection of resources such as fisheries, etc.

However, classical liberals are also concerned about the growth of international law. In most Western judicial systems, treaties and comparable international agreements are of a higher order, which means they overrule national law. Therefore, governments have to be careful with what they agree to in international negotiations since this impacts on the lives of their citizens. It may be tortuous to reach agreement in the international arena, but it is even harder to change or abolish existing treaties. Mises and Hayek protested against the explosion of international law, especially after World War II. The new international laws entailed too many positive rights and claims (rights to), instead of the negative rights (protections against) favoured by classical liberals. To them, this amounted to constructivism at the international level. An example is the Council of Europe's European Convention on Human Rights (ECHR), which has a direct effect on national penal systems and many other domestic judicial arrangements.

As an aside, even though the UK has left the EU, it has not left the ECHR, since it remains a member of the Council of Europe, which consists of 47 member states compared to the EU's 27. In fact, the UK joined the Council of Europe 24 years before it joined the EU (Equality and Human Rights Commission 2017).

International governmental organisations (IGOs)

IGOs are constituted by treaties, so the above also applies to them. This is problematic for classical liberals, as some have become state-like structures at the international level. Again, there is no single classical liberal rule here, because there are so many different IGOs. Some tasks may be approved while others clearly are not. As a rule of thumb, those IGOs that cover technical aspects of international cooperation tend to make sense, even though there is sometimes political debate involved in agreeing on technical issues. Many of the well-known large IGOs meet classical liberal resistance, for example (parts of) the United Nations, the World Bank and the World Trade Organization. Again, specific case studies are needed to determine this for each international organisation.

Based on their experience with the failed interwar League of Nations, Mises and Hayek were among the first to criticise the UN (see Van de Haar 2009). They thought the UN was constructivist and a poor compromise between the West and the Soviet bloc after World War II. They were concerned that some social and economic (positive) rights were included in the Universal Declaration on Human Rights. On top of that, these rights were given practical meaning by the extensive system of the UN's daughter organisations. Yet, as indicated, this must not be seen as a complete rejection of these organisations. For example, the World Health Organization (WHO) is active in a field where Hayek and Mises considered government action justified. Most of the work of the WHO is technical: it is about exchanging information, prevention of and fighting disease. There are of course political debates within and about the WHO, especially given its poor response to Covid-19 and its perceived closeness to the Chinese government, which has refused to be open about where and how the virus originated. However, in theory, it should have a role in preparing for pandemics and other health-related issues with an international character, provided the same could or would not be done by private organisations such as the Red Cross.

Without going into detail, classical liberals remain sceptical of many organisations within the UN system, including the core of the UN itself. The General Assembly may serve some purpose as a meeting place for state representatives discussing international issues. Yet the Security Council is a clear anachronism, especially the Permanent Members, since they represent post–World War II power relations, with China taking the place of Taiwan in 1971. This is not necessarily representative of power relations in the world today, or may foster power relations that are clearly anachronistic, while the UN has not always covered itself in glory in dealing with conflicts around the globe, to put it mildly.

Classical liberals want to get rid of those IGOs that interfere with free markets, such as the International Labour Organization, the World Bank and possibly also the International Monetary Fund. The WTO is also criticised, since free trade does not require so much state interference as to warrant an international organisation, although it is also important to have an IGO whose remit is to promote open trade.

In the same fashion, there are pros and cons considering the EU, partly due to its uniqueness in being a constantly evolving entity. It has played a key role in keeping peace in Europe, although this peace was also fostered by the Cold War balance of power and NATO. The EU is far from the Hayekian ideal of a federation. It does not pool important powers at the federal level (defence, foreign policy), while many of its policies (agricultural, economic, monetary, structural, regional, etc.) have no place in a classical liberal ideal at all. On the other hand, a more positive assessment is possible for the internal market (although some rules and regulations require closer scrutiny) and the Schengen area, which provides free movement of people. The common rules for asylum, immigration and border patrol are a good idea in principle, even if they have been ineffective in practice. While the EU's trajectory towards a superstate poses threats to individual liberty, it also has classical liberal features.

Without going into detail about other IGOs, it should be noted that IGOs should comply with the classical liberal principles of freedom. IGOs are not inherently good or bad. They might serve useful goals, but they could just as well be a hindrance to international order and/or individual liberty. Critical scrutiny remains important.

International Non-Governmental Organisations (INGOs)

Recent decades have seen an increase in the activity and awareness of INGOs. Classical liberals have no objection to them in principle since they are a result of freedom of association. While classical liberals disagree with many of the goals of INGOs, for example, those that campaign for trade protectionism or increased state intervention, there are other INGOs that favour classical liberal policies. Classical liberals do object to the common practice of governments funding NGOs with taxpayers' money in the belief that they can achieve policy goals that governments cannot. NGOs receiving taxpayer funding from governments cannot truly be considered as 'non-governmental' and in fact lay themselves open to the accusation of being an extension of governments. The relationship sometimes becomes murky when INGOs lobby governments after receiving taxpayer money from those governments. In this sense, it is a clear example of how policy outcomes are often the result of lobbying by interest groups, rather than a government acting in some supposedly neutral way in the public interest. For classical liberals, these practices are another reason to favour a limited state that funds its core roles rather than funding special interests.

9 ECONOMICS: TRADE, GLOBALISATION AND DEVELOPMENT AID

The introduction of the building blocks of the classical liberal theory of international relations finishes with the topic that classical liberals are best known for in the debate about international politics. How do international relations deal with the element of economics? Besides free trade, and its relation to peace, this chapter deals with globalisation and development aid.

Economists point out that scarcity, defined as resources not being endless, has an impact on human life. This is no different for international life. International economics, specifically trade, are in fact one of the oldest forms of international cooperation. Classical liberalism has long been associated with support for free trade. Contemporary classical liberals continue to support this agenda but have broadened it to include support for globalisation. However, bad economics also exist at the international level.

Free trade

Classical liberals have supported free trade for centuries. As long as people have inhabited the earth there has been trade. In that way 'the propensity to truck, barter and exchange' is really part of human nature, as Adam Smith argued. This fundamental idea has been confirmed by empirical evidence time and again. International trade ensures a more efficient distribution of resources, goods and services, through specialisation, separation of labour and the innovation that comes with it. While there may be some who lose out in the short term, free trade tends to make people wealthier. Not only the rich people, but everyone (Panagariya 2019). Trade makes the 'economic pie' bigger, making everyone richer. This also applies to countries that may appear to have no absolute advantages and hence fewer possibilities for trade. As Ricardo (2002) pointed out, they still have a comparative advantage, which allows them and their trade partners to gain from such trade. In addition to these economic gains, trade also fosters the exchange of ideas and cultural experiences, as Hume (1987) argued in his essay *On the Jealousy of Trade*. These fundamental ideas and their positive outcomes are still valid.

In liberal eyes, free trade is best when it truly is free. That means trade between individuals (or individual companies), wherever in the world, without the interference of governments or international organisations. This has been hard to accomplish. While the EU's internal market has come a long way, there are still barriers to trade within the EU, and the EU's external trade policy is seen as open in some areas while protectionist in others. The US is no better in this respect, while most countries in the world are far worse.

History shows that attitudes towards international trade are constantly changing (Irwin 1996). Sometimes it is popular and embraced by economic and political elites, but there are always counter-movements when free trade is curtailed due to a rise in nationalist and protectionist sentiments, influenced by non-liberal thinkers. Governments have almost always seen reasons to limit, regulate or tax trade through tariffs, quotas or regulations. Irwin explains that the Ancient Greeks and Romans looked down on trade, as did many early Christian Church leaders. It was not until the eighteenth century that the idea of free trade became firmly anchored within the classical liberal tradition when Smith (1981) wrote *An Inquiry into the Nature and Causes of the Wealth of Nations* in 1776. However, we should not view Smith as the 'inventor' of free trade ideas. Beside his own contributions, he built on and improved those of the French Physiocrats, the Spanish School of Salamanca, Dutch thinkers (for example, Grotius and Bernard Mandeville), English and Scottish thinkers, as well as the fourteenth-century North African scholar Ibn Khaldun.

The nineteenth century saw perhaps the most well-known battle over trade, started by the activists of the Manchester School, most notably Richard Cobden and John Bright. They fought for a repeal of the protectionist Corn Laws, and they ultimately succeeded. In 1860, Cobden also negotiated an Anglo-French trade treaty (the Cobden–Chevalier Treaty). However, these ideas were challenged, among others, by German thinker Friedrich List (1841), who argued for regulations and restrictions on trade.

His ideas have remained an inspiration for protection-ist thinkers ever since.

The tide appeared to turn when, after World War II, the tariffs between the industrialised nations were negotiated downwards under successive General Agree-ment on Tariffs and Trade (GATT) rounds. An Inter-national Trade Organization (ITO) had been proposed at the Bretton Woods conference in 1944, which resulted in the establishment of both the World Bank and the International Monetary Fund (IMF). Since the ITO was never ratified by the American Congress, international trade negotiations continued under GATT until the es-tablishment of the World Trade Organization (WTO) on 1 January 1995.

The WTO included a dispute settlement process and also became the forum for negotiations to reduce non-tariff barriers. As WTO membership expanded and the agenda broadened, the many different interest groups and protests led to the effective abandonment of the Doha Round of multilateral trade negotiations, which had begun in 2001. While small steps have been taken, most countries have resorted to negotiating bilateral or plurilateral trade agreements. While opponents see the WTO as an example of a liberal world order, Sally (2008) argues that impulses for freer trade do not come from international institutions. Classical liberals regard the WTO as a second-best solu-tion to truly open trade, where most or all trade barriers are removed. So-called free trade agreements (FTAs) are in reality preferential trade agreements (PTAs), where many sectors remain protected from international competition.

The WTO and PTAs should therefore be seen as examples of regulated trade as opposed to true free trade.

Free trade and peace

Many people in the liberal tradition have overly optimistic ideas about the pacifying effects of trade. Yet there is no inherent relationship between trade and peace (Van de Haar 2010, 2020a). The predominant idea underlying the trade-fosters-peace thesis is as follows: the buying and selling of goods and services across borders is a peaceful activity. Agreeing on trade deals fosters ties between the individuals involved on both sides of the border. All those individual ties from trade add up, foremost economically, but also socially and politically. This makes states interdependent. As a consequence, when contemplating war, leaders of countries also have to take into account the harm done to trade relations between countries and the economic costs involved. Allegedly, they want to avoid that, so in this way, trade fosters peace. While the idea that free trade leads to peace is a seductive one, the world and our choices about making war are not that simple.

David Hume and Adam Smith had already noted that there is no automatic relationship between free trade and peace. For one thing, human nature is still unchanged, so violence and conflict still exist, no matter how much trade there is. Also, even though international trade is between willing buyers and sellers in one country trading with their counterparts in other countries, trade is seen to make certain states richer, which leads to what Hume called 'the jealousy of trade.' The governments of richer nations may use their increased tax revenue to purchase more weaponry, which may lead to an arms race and more conflict-prone foreign policy.

Trade is unable to foster peace because it is unable to overcome the many other causes of war, such as cultural and religious differences, the fight for natural resources, including – increasingly – rare materials, or more traditional conflicts between great powers or their proxies over border disputes. States may also act against their economic interests for some perceived higher goal, as war specialist Coker (2014) explained. The literature points out that the causes of war are often multifaceted and complex. Wars happen because people have reasons in the form of goals and grievances, enough resources and the resolve to wage them. Trade relations are just one factor in the mix of causes of war, let alone such coincidental issues as chance, luck or reckless behaviour by individuals who happen to influence public policy. Suganami (1996) argues that international commerce is simply not a 'perfectly effective anti-war device.' The most free traders can hope for is that the protection of trade relations may be one of the factors in a government deciding not to wage war. Sometimes trade does foster peace, but it is just one of the elements in the consideration. For example, Ukraine and Russia still had trade relations in 2022, even though these had significantly decreased after Russia took the Crimea in 2014. It did not deter Russia from invading Ukraine.

The debate on the relationship between trade and peace is not the same as the more recent 'democratic peace' debate, sometimes also referred to as the debate on 'liberal peace'. This is the empirically more robust claim that settled democratic countries do not fight each other, although they may regularly wage war with other countries or fight each other when they are still young and unstable. The search for an empirically robust explanation has taken international relations scholars over two decades without reaching a firm conclusion, partly due to methodological constraints. Sometimes one of the variables in this kind of research is trade, although this mostly concerns bilateral trade while multilateral trade and intra-firm trade are often overlooked. At the time of writing, the debate continues.

Globalisation

Just because trade in itself may not always lead to peace, it does not follow that the other positive effects of trade should be overlooked. Classical liberals argue strongly that the expansion of free trade is a good thing and therefore support further liberalisation. Through international trade, but also as a result of modern communications, reduced travel times, instant news media, financial markets and other innovations, the world has become more entwined and interdependent than ever before. It almost goes without saying that today's world is much more globalised than ever before.

While there are many definitions, globalisation is defined here as the global phenomenon of increasing interconnectedness of people, through economics, sports, communications, travel and cultural exchange. Globalisation is a classical liberal ideal, because it increases the possibilities for individuals to live their lives in greater liberty, depending of course on local or specific circumstances. On average, a young Western woman will benefit more from globalisation than an old person in Myanmar or Afghanistan. But all people still benefit, perhaps even except those suffering from the harshest and secluded dictatorial regimes.

There has been a multifaceted debate about globalisation for at least two decades now. Some people argue that trade and globalisation are bad for the environment, there are trade unionists who focus on the groups who find themselves outcompeted in a globalised economy, and some who argue that the developed world is getting richer at the expense of developing countries, in terms of income but also through the exploitation of people in the poorer countries by Western multinational companies, including child labour. And these are just a few examples. However, Bhagwati (2004) claims that most of these arguments are either false or that the problems are partly or wholly due to domestic policies in the countries concerned. Many opponents of today's globalisation are likely to be anti-capitalist, anti-corporate or anti-West. Indeed, Marxists argue that there have been earlier episodes of globalisation connecting different parts of the world based on social, cultural and political exchanges, but that contemporary globalisation should be viewed as global capitalism (Germann 2018). Norberg (2001) appears to acknowledge this point in the title of his book *In Defence of Global Capitalism,* while

Wolf (2005) defends today's globalisation in his book *Why Globalization Works*.

In formulating a classical liberal theory of international relations, there will be questions about the relationship between globalisation and the nation state. For example, does globalisation erode the ties between individuals and their nations in favour of a cosmopolitan citizenship? This remains a debate within liberalism today. Globalisation is about freeing up the world economy, increasing opportunities for trade as well as addressing issues such as increasing communications and social, political and cultural exchange. It also works: the world is becoming a better, cleaner and healthier place on many accounts, which is certainly due to capitalism and globalisation (Norberg 2017; Rosling 2018). For many classical liberals, these are links that do not necessarily need state governments to be involved. However, many aspects of globalisation do require regulations or agreements between states, so in many ways, a globalised world can also be viewed as a world of states. Classical liberals defend and want to expand globalisation, in the interest of individual liberty and well-being, all over the world.

Development aid

Classical liberals think free trade and globalisation are the best ways to contribute to the development of everyone, including the least well-off. Yet prosperity depends on good domestic policies and institutions, such as the protection of property rights and an honest and accessible judiciary.

Many classical liberals have opposed government-to-government development aid. The classical liberal maxim is 'trade, not aid'.

Development economist Peter Bauer (1971) was among the first to point out the unintended consequences of development aid, claiming it was 'bringing money from the poor of the rich countries, to the rich of the poor countries'. Government-to-government development aid in the form of money transfers has often ensured that elites in developing countries stay in power while the fate of the wider population in recipient countries has hardly ever improved. Classical liberals and others argue that poorer countries have a better chance of becoming wealthier if they introduce better domestic policies such as the eradication of planned economies, the liberalisation of social life, and increased trade with each other and with developed economies. Development depends on people themselves. They must be allowed the opportunity to improve their life circumstances.

Bauer was a friend of Ludwig von Mises and F. A. Hayek, who also strongly opposed state-to-state development aid. Mises argued that only capitalism and laissez-faire policies could help improve poor people's circumstances. The West did not have to feel guilty for countries (often former colonies) that failed to do so because, in many cases, it was their collectivist economic policies that led to poverty. Hayek fully supported this line of thought and pointed out that the leaders of newly born countries often got their socialist ideas from the Western universities where they were educated. Bad policies would keep countries poor. An example was the far too rapid industrialisation that many countries instigated. This was detrimental to the regular development of agricultural policies, which would have been able to deal with shortages in the food chain. The general position of Bauer, Mises and Hayek was that without capitalism these countries would only become poorer (Van de Haar 2009).

These classical liberal insights have become increasingly popular due to empirical research conducted by a number of economists, such as Easterly (2002, 2013), and Zambian author Moyo (2009), who declared herself a disciple of Peter Bauer. Recent Nobel Prize winners Banerjee and Duflo (2011) would probably not consider themselves classical liberals, yet their empirical research into the daily lives and conduct of poorer people often offers evidence with which classical liberals agree. Palmer and Warner (2022) emphasise that prosperity can only be achieved if individuals are valued as self-governing agents and a 'dignity-first' approach to development is taken.

While classical liberals are sceptical of governmental development aid in the form of bilateral and multilateral gifts, they are not against non-state charity, believing that non-governmental organisations (NGOs) should remain free to collect money and spend it in the way they, their donors and recipients see fit. The aims and effectiveness are a matter for them, though classical liberals may disagree with the policies advocated by a number of international NGOs, who often have a paternalistic view and sometimes continue to advocate trade protectionism and other socialist policies. The need for immediate foreign emergency relief after a natural disaster or conflict is less controversial, but policies around reconstruction can sometimes lead to governments increasing their controls and restricting economic freedom. On the whole, classical liberals are convinced that while some governmental development aid may sometimes reach the intended recipients and those most in need, domestic policies and people-to-people links are far more important.

Putting the building blocks together

The most important building blocks of the classical liberal theory of international relations have now been introduced. Again, much more can and often has been said about each of them. Yet the goal here is to present the main contours of the theory, not to discuss every detail. Also, as was underlined in the introduction, a theory is never complete and cannot tackle all possible questions or issues.

However, compared to other theories of IR, the classical liberal theory

- originates from first principles;
- has a solid foundation in the ideas and theories of its most important representatives, with sufficient overlaps over a period of centuries;
- includes more concepts than most other (liberal) IR theories;
- is able to help analyse, and provide answers and solutions to many questions of international relations.

Classical liberal IR is a bottom-up theory, based on a realistic view on human nature and the relationship between individuals and groups, particularly the nation and the state. It is natural for many classical liberals to regard the nation state as the main actor in international relations. Classical liberalism recognises the inevitability of conflict and war and aims to deal with their regrettable occurrences. On the one hand, this can be achieved through spontaneous ordering forces, such as the balance of power, while on the other, some limited but strong international law is needed to protect classical human rights, together with a limited number of functional international governmental organisations. Sound capitalistic principles must be applied, such as free trade (which, while not fostering peace, brings other enormous benefits), globalisation and a scepticism of government-to-government development aid. Lastly, there is the demand for restraint: military intervention as a last resort, some limits on immigration (although this remains a point of classical liberal debate), no place for imperialism, let alone collectivist notions such as nationalism.

Summing up, the morphology of the classical liberal theory of international relations can be seen in table 3.

Table 3   A morphology of classical liberal international theory

Core concepts	Realistic view of human nature; nation; state; sovereignty; just war; balance of power; free trade; globalisation.
Adjacent concepts	Human rights; diplomacy; international law; international organisations.
Peripheral concepts	Immigration; military intervention; federation (as a last resort).

It is important to note is that there is no place for empire, nationalism and government-to-government development aid.

PART III: INTERNATIONAL RELATIONS THEORY

10 LIBERAL IR THEORIES

Surprisingly, given the widely acknowledged influence of liberalism in IR, there is little literature on this topic, and even less on classical liberalism and IR theory. In his recent edited book, Jørgensen (2021) is one of the few academics who has attempted to provide a full overview, albeit just from a European perspective. In this part, we try to provide more context. First, we examine the relation between the classical liberal IR theory and the other liberal theories of international relations. The next chapter will zoom in on the question of how to see classical liberal IR within a broader IR theoretical framework, while the third chapter in this part will present the views of libertarians on questions of international relations. The overall aim of this third part is to show where classical liberalism fits into the overall liberal IR framework.

After a general introduction, this chapter provides a brief overview of the main liberal IR theories:

- liberal internationalism;
- (neo) liberal institutionalism;
- functionalism and interdependence;
- regime theory;

- embedded liberalism;
- liberal peace theory;
- other liberalisms.

The contours of liberal IR theory

Zacher and Matthew (1995) describe liberalism as progressive, cooperative and modernist, which is the opposite of realism, for a long time the other dominant paradigm in international relations theory. The label 'progressive', related to social liberalism introduced in the first part of this book, means, according to Zacher and Matthew, that liberals believe that their approach to international relations will foster greater human freedom by establishing peace, prosperity and justice, based on the power and use of human reason. Liberalism is also seen as 'cooperative', since it emphasises cooperation between states and other international actors. It is 'modernist', since liberals are seen to believe that international relations can be transformed by the embrace of concepts such as liberal democracy, international interdependence and institutions, as well as the integration of groups and individuals, to modernise the world. Zacher and Matthew argue that these three elements capture the main liberal ideas about world politics, although they also distinguish different liberal IR theories. For them, liberalism stands for the dramatic improvement of the world in terms of moral character and material welfare. These are dramatic claims, which liberal IR theorists often repeat in one form or another. Therefore, it is understandable that liberals are often called idealists, or utopians.

Compared to the classical liberal theory presented here, most other liberal theories have less solid groundings in political philosophy. Only a few draw explicitly on the writings of liberal political theorists, while others adopt just one or more liberal ideas and use them for their 'liberal' theory. The result is unimpressive, to say the least. In her book on liberal internationalism, Jahn (2013) notes that liberalism in IR theory is fragmented, diverse and poorly or at best partly defined, while a clear relationship to liberal political theory is often lacking.

For example, as indicated before, many current liberal IR theories are roughly social liberal, which leads to a concern with social justice. This wholly or partly draws on the work of John Rawls and his seminal work, *A Theory of Justice* (Rawls 1999), which is seen as one of the most influential books on political philosophy in the twentieth century. In it, he defined justice as fairness and believed that a just society should protect classical human rights but also comply with two rules:

(1) equal chances for all;
(2) the progress of richer people is only justified when poor people benefit as much.

Yet this was only meant for politics within bordered societies. In *The Law of Peoples* (Rawls 2002), he considered the implications of his theory for international affairs. He concluded – much to the disappointment of many of his intellectual admirers – that his earlier theory was not applicable internationally, therefore he did not believe in a cosmopolitan world order of just states. Instead, he concluded that only a small group of states could ever be just. As a consequence, his international theory focused on how the few just states would have to deal with the larger and more diverse group of unjust states, who viewed justice in a different way. Rawls considered international inequality a given, and it is important to note that Rawls draws a distinction between justice in domestic politics and in international politics. Not all modern liberal IR theories make the same distinction.

Nevertheless, to provide a bit more detail than Zacher and Matthew, the common threads running through liberal IR theories are the following, including a belief that:

- world peace is attainable, since humans are seen as rational enough to overcome war and conflict;
- the nation state is a problematic actor in world affairs, which needs to be curtailed, especially the alleged 'warmongering' elites, such as diplomats, military commanders and/or the so-called military-industrial complex;
- there is an alleged pacifying influence of interest groups and public opinion on foreign policy decision-making;
- peace-oriented foreign policies can also be fostered by domestic institutional arrangements, most notably democracy;
- there is an important role for intergovernmental and non-governmental organisations, regimes and international law to overcome or neutralise the effects

of power politics and to take power away from the national state;

- international trade helps foster peace;
- and more recently, support for humanitarian intervention to promote democracy.

In short, it is about working together internationally, using international cooperation to diminish the role of the state and foster peace, enhancing regimes domestically. Indeed, a rather optimistic world view.

Liberal internationalism

The oldest liberal theory of international relations is liberal internationalism. It aimed not only to describe international relations but to reform the world in order to make it more peaceful. Liberal internationalists have a rather positive view of human nature, unlike classical liberals. They think highly of the possibility of bringing lasting change to the international realm through state action at the international level.

The origins of liberal internationalism date back to the eighteenth and nineteenth centuries. In Britain, it can be traced back to Cobden's appeal for free trade and non-interventionism, the Benthamite emphasis on international law and Kant's transnational interstate organisation.

While there are many different liberal international theorists, all concentrating on their own variations, they have the following ideas in common, which are almost identical to the most important characteristics of liberal

IR theory mentioned above. This says a lot about the enduring influence of liberal internationalism. In addition:

- Liberal internationalists criticise diplomacy, war and the balance of power, and aim to circumvent or eradicate them.
- International reform must be accomplished through domestic reform, by decreasing the influence of diplomats and generals, by increasing the influence of the alleged peaceful general public, and, in the last decades, by fostering democracies.
- The establishment of international organisations is helpful, as is the design and enforcement of international law, rules and norms. Liberal internationalists believe that international organisations help to overcome international anarchy and promote international justice.
- Enlarging international (economic) dependencies will stimulate peaceful solutions to conflicts in international relations.

The oldest international organisations were founded in the second half of the nineteenth century to foster international cooperation on mainly technical issues. The first one was the International Telegraph Union (ITU), which was founded in 1865 and later became the International Telecommunication Union. Liberal internationalists also put a lot of emphasis on international arbitrage to settle disputes, while the early twentieth century saw more emphasis on the development of international law, for example, through a series of peace conferences at The Hague, where the International Peace Palace would be built, largely paid for by American steel magnate and philanthropist Andrew Carnegie.

Some IR theorists view US President Woodrow Wilson's Fourteen Points speech in 1918 as the pinnacle of liberal internationalism. In this speech, President Wilson outlined his ideas for a post–World War I world, demanding 'open covenants of peace openly arrived at, free international economic transactions, freedom of the seas, the reduction of national armaments, self-determination for colonised people, and the formation of a general association of nations, for the purpose of guaranteeing political independence and territorial integrity to great and small states alike' (Wilson 1918). This led to the founding of the League of Nations after World War I, which was seen as a possible antidote against future war, in line with the thinking of academic IR departments at the time.

Much more can be said about the history of liberal internationalism, of which Jørgensen (2018, 2021) provides a balanced critique. He concludes that liberal internationalism has been a complete failure since its main goal, the avoidance of conflict and war, has never been remotely within reach. On the other hand, liberal internationalist ideas have remained influential. For example, although the League of Nations was generally seen as a failure, after World War II it was succeeded by the United Nations (UN), complemented by the international organisations of the Bretton Woods system, and since then by the emergence of many other international organisations and regimes.

Liberal institutionalism

In many ways, liberal institutionalism, sometimes referred to as neoliberal institutionalism, is a modern variant of liberal internationalism. The main difference is that it recognises the central role that states play in international relations, although it attempts to mitigate the alleged negative behaviour of states through intergovernmental institutions and international organisations, which are seen as promoting international order and peace. Liberal institutionalists tend to focus on the main international organisations, such as the United Nations, the World Bank and International Monetary Fund, the World Trade Organization, but also on several important treaties, including the Nuclear Non-Proliferation Treaty, or regional cooperation in the Association of Southeast Asian Nations, the Gulf Cooperation Council, and so on.

The prefix 'neo', or the term 'neoliberal institutionalism', refers to a newer, more positivist methodological approach, yet it is debatable whether there is much difference between liberal institutionalism and neoliberal institutionalism (see Stein 2008). As an aside, realism also experienced a reincarnation into neorealism, which led to the so-called 'neo-neo' debate that dominated IR theory for a number of years.

Liberal institutionalists also explore the possibilities for states to cooperate in international regimes and institutions. The language they use is more connected to their attempts to bring the methods of the natural sciences into IR. For example, they talk of absolute and relative gains (taken from game theory) for governments in their dealings with international institutions. This is not just talk about theory. Goddard and Krebs (2021) argue that liberal internationalism and liberal institutionalism have been key features of US foreign policy to legitimise American power after the end of the Cold War, when America was the sole global superpower for at least two decades.

Functionalism and interdependence

Some liberal institutionalists, most notably in Europe, pursue increasing supranational integration. This is the idea that if states pool their sovereignty and resources into regional or even global organisations, this may lead to peace and prosperity. An early version of this, known as functionalism, was developed by David Mitrany (Ashworth 1999). He contended that international reform depended on transnational associations instead of interstate mechanisms to create technocratic, function-specific international organisations that could foster international planning.

Functionalism can be viewed as a variant of the idea of international interdependence, examined earlier in relation to free trade and peace. In *Power and Interdependence*, Keohane and Nye (1989) detailed how complex interdependence works. They suggest that individuals, groups and states get entangled in a series of political, economic and social relations through multiple channels, resulting in a situation of complex interdependence. As a result, they believed that military relations would become less important and the costs of conflict would be much higher.

Regime theory

A related concept is regime theory, which mainly focuses on social institutions that govern specific topics of international cooperation through formal and informal rules and agreements. These regimes are seen as specialised arrangements, which cover a wide spectrum of functional fields, ranging from armaments control or fisheries agreements to specific parts of the environmental agenda, such as tackling deforestation or the use of pesticides. As Young (1989) argued, it is quite significant if states are willing to co-operate on a global scale to solve collective action problems. Regimes are easier to establish than, for example, world government or other forms of global public authorities. Some regimes cover the whole world, while others cover smaller geographical areas. Some have two or three members while others have many more. Sometimes these members are just states; in other instances they also include non-state actors. Regime theorists argue that regimes make the international arena more predictable and less anarchic. Classical liberals do not object to these types of rules and agreements or the spontaneous formation of all kinds of multiple channels as long as they benefit individual liberty and do not turn into top-down international organisations seeking to coerce individuals or state governments.

Embedded liberalism

Another variant of liberal institutionalism is Ruggie's 'embedded liberalism' (1982), which argues that liberalism was secured in the post–World War II US-led international order through international organisations such as the International Monetary Fund and the World Bank. These were 'embedded' in domestic arrangements where states were seen to be able to 'soften' the alleged hard effects of the laissez-faire international economic system with interventions such as welfare states and other social arrangements. Ruggie feared that states would continue to lose their capacity to intervene to embed liberalism due to globalisation and the resulting international policy competition between states.

The empirical validity of Ruggie's claims is questionable. The assumed laissez-faire economic system is nowhere to be seen. In fact, the state sector in many countries was between 40 and 50 per cent of gross domestic product (GDP), before the Covid pandemic. Many of the supposed 'softening measures' have arguably done much harm to individual freedom and prosperity (see Bernstein and Pauly 2007). The World Bank is certainly not a classical liberal institution, but rather a collectivist development aid bureaucracy, whereas the IMF assisted many non-liberal states and their elites survive, despite the alleged 'harsh conditions' attached to its loans in its role as a last-resort lender. Therefore, classical liberals will question both the empirical basis of 'embedded liberalism' and how liberal the theory was anyway.

Liberal peace theory

One of the important debates in IR theory over the past decades has been whether democracies go to war with each other or not. This also goes back to those theorists who point out the importance of domestic arrangements for international order and peace. While there has been no definitive verdict, there is an empirically robust claim for the peace that exists between a number of developed countries post–World War II (Brown et al. 1996).

The interesting point here is that democratic peace is often referred to as 'liberal peace', allegedly due to the existence of liberalism in the countries concerned. Owen (1997) and MacMillan (1998) are good examples. They use the term 'liberal' rather loosely with reference to a limited number of liberal characteristics. An example is where one liberal thinker – most often Kant – is taken as the standard for liberal thought and the 'republican peace', based on a poor reading of his book *Perpetual Peace* (1795), overlooking other views as well as ignoring evidence that questions Kant's position as a liberal standard bearer. Molloy (2017) offers a good critique in his book *Kant's International Relations*.

No doubt, the term 'liberalism' is often used loosely in liberal peace theory. This is not to argue that all the elements or ideas used are illiberal, but that the selected liberal theories do not present a complete enough picture. Many classical liberals have their doubts about democratic peace theory, since it is based on an optimistic view of human nature, the capability of domestic arrangements to overcome fundamental traits in the international system, and the peace-enhancing effects of trade and other policies. Despite an enormous amount of research over the past few decades, there is still no robust explanation for democratic or liberal peace theory.

As referred to in part II, the same goes for the idea of 'capitalist peace', which remains a topic of debate in the IR literature (see, for example, Mansfield 1994; Barbieri 2005; Mansfield and Pollins 2003; Schneider and Gleditsch 2013; Copeland 2015). Some writers attempt to use historical ideas in this contemporary debate, claiming to present 'classical liberal ideas'. Gartzke, in his influential article 'The capitalist peace' (2007), introduces many writers, yet it is hard to count them among the classical liberal founding fathers discussed in this book. For example, his focus is on Cobden, but also on Mill, who was a mainly social liberal, or on thinkers without any classical liberal credentials, such as Rousseau, Bentham and Norman Angell. McDonald (2009) is not much better, picking and choosing his way through the history of ideas, sometimes stepping on a classical liberal idea, yet without any solid analysis.

Other liberalisms

As in political philosophy, where there exists a number of different liberalisms, including natural law liberalism, economic liberalism, progressive liberalism, etc., IR theorists also attach a broad array of adjectives to the word 'liberalism'. Many of these labels are often rebadged or minor variations of existing liberal ideas and theories. These are often 'invented liberalisms', in the sense that the author cannot resist the temptation to introduce her own kind of liberal variation. Here we will refer to them as 'other liberalisms'.

Michael Doyle and Robert Keohane have been particularly influential in rebadging or recasting existing liberalisms. We have already seen commercial pacifism (sometimes referred to as commercial liberalism) and republican liberalism. Sociological liberalism means to describe transnational interaction and integration, not unlike functionalism. Interdependence liberalism also refers to functionalism and liberal institutionalism. Keohane also introduced 'sophisticated liberalism', which is a combination of commercial and republican liberalism.

Zacher and Matthew (1995) suggest a couple of other liberalisms, including military liberalism, based on a mutual interest in peace due to the deadly power of modern military technology, and cognitive liberalism, based on the idea that rational behaviour and increased knowledge lead to more peaceful international relations.

In a critique of American foreign policy aimed at 'liberal hegemony', Mearsheimer discusses the debate or battle between 'progressive liberalism' and 'modus vivendi liberalism' (Sinha 2017). The former requires government intervention, while the latter argues that rights are about freedom to act without government interference. This is not unlike the difference between classical and social liberalism. Rengger (2013) writes about 'dystopic liberalism', or the liberal variant underpinning Rawls's *Theory of Justice*, which could be seen as social liberalism.

Griffiths (2011) further muddies the water by arguing that liberal internationalism has three main pillars: republican liberalism, commercial liberalism and regulatory liberalism. They largely comply with the characteristics listed above. Republican liberalism is about spreading democracy, based on the idea that developed democracies do not fight each other. Commercial liberalism is a repetition of the 'trade leads to peace' thesis, while regulatory liberalism is less about regulation and more about international rules.

There are exceptions. Richardson (2001) appears to grasp the differences between social and classical liberal ideas on international relations, although his antipathy towards the classical variant remains unsubstantiated. Doyle (1996), in other works, recognises the differences between liberalisms in foreign policy, pointing out that classical liberalism leans towards power politics, while social liberalism aims at pacifist global justice. These authors still provide a partial analysis, but at least they provide their readers with some in-depth analysis of liberal thought.

Summing up, the shared characteristics of the numerous liberalisms in current IR theory are that they are largely optimistic, believing in the possibility of making the world a better and more peaceful place, based on a harmony of interests and international cooperation leading to a decrease in war and conflict. To achieve these ends, international organisations, law and regimes are promoted. Liberals also focus on changing the nature of domestic politics via the power of public opinion, which they see as inherently peaceful. These warm, almost idealistic beliefs are not shared by classical liberal international relations theory, as we saw in the previous part and as the next chapter again makes very clear.

11 CLASSICAL LIBERALISM, IR THEORY AND FOREIGN POLICY

Given that most contemporary liberal IR theories differ from classical liberal ideas, how should we view a classical liberal IR theory within a broader IR theoretical framework? Classical liberalism is not only a philosophical lens through which to view international relations but also a normative theory. It is how classical liberals think we should look at world politics and how international politics should be conducted. This raises the question of how this could be achieved. To answer this question, a set of general principles of a classical liberal foreign policy will be considered.

Classical liberal versus liberal IR theory

The previous chapters examined the differences between classical liberal IR theory and other liberal IR theories. These are summarised in table 4.

These differences between classical liberalism and current liberal IR theories at least show that there is a need for much more intra-liberal debate on international relations, just as there is a debate among liberals on domestic politics. However, this will also require the recognition of a distinctly classical liberal IR theory in academia as well as in foreign policy circles.

Table 4   The differences between liberal and classical liberal IR theory

Liberal IR theories	Classical liberal IR theory
World peace is attainable, since humans are seen as rational enough to overcome war and conflict.	Conflict and war are viewed as perpetual characteristics of international relations, based on a realistic view of human nature.
The nation is seen as a problematic actor in world affairs.	The nation is the prime and a natural actor in international relations.
Balance of power is problematic and a cause of war.	Balance of power is a spontaneous ordering mechanism, which fosters international order.
Other actors: the alleged influence of 'warmongering' diplomats, special interests and the so-called military-industrial complex need to be curtailed. The opinions of NGOs and the public are important factors to take into account in foreign policy decision-making.	Other actors: There is a neutral view on the role of diplomats, interest groups, NGOs, public opinion and other internal actors.
The full catalogue of human rights needs to be defended.	Only classical human rights need to be defended.
Peaceful international relations can be fostered by domestic institutional arrangements, most notably democracy (democratic peace theory).	Sceptical about the ability of domestic arrangements to overcome conflict and war.
Important role for intergovernmental and non-governmental organisations, regimes and international law, which aim to overcome or neutralise the effects of the logic of power politics.	The role of international law, regimes and intergovernmental organisations is important but should be limited and mostly functional since they can become a threat to individual liberty.
International trade is expected to foster peace.	International trade does not inherently lead to peace, although it is a very positive thing and can foster good relations.
Fairly broad support for military intervention, also for democracy promotion.	Military intervention is only acceptable in exceptional instances, such as genocides. No belief in success of democracy promotion.

Classical liberalism and the English School of international relations

So far, classical liberalism has been compared to other liberal IR theories and realism. While a comparison between classical liberalism and other IR theories is interesting, it is clear that classical liberals disagree with IR theories such as Marxism, Critical Theory, Constructivism or Green Theory. More promising is to see how classical liberalism relates to the English School of international relations, which was developed by Hedley Bull, Martin Wight, Herbert Butterfield, Adam Watson and other members of the British Committee on the Theory of International Politics, in the decades after World War II. In the past twenty years or so, English School theory has become more influential and is often a part of IR theory courses.

The English School is sometimes equated with a Grotian or international society approach. Initially, the novelty of the English School was the emphasis it placed on the idea of an international society as a middle way between the two leading paradigms at the time: realism and liberalism. Linklater and Suganami (2006) note that a society of states exemplified progress from a situation of a system of states (explained below). Another characteristic of the English School theory is the use of three traditions in IR theory to analyse world politics (Wight 1991). The three traditions are known by different labels: Hobbesian, Grotian, Kantian; or Realism, Rationalism, Revolutionism; or international system, international society, world society. The traditions are useful as methodological aids, but should not be seen as 'railroad tracks running parallel into infinity', as Wight put it. There is some overlap between them, and not all ideas can be neatly fitted under one label.

While attractive, the use of the Hobbesian, Grotian and Kantian labels has been found inadequate since the ideas that comprise the three traditions do not necessarily match the ideas of these three thinkers. In fact, there is a whole sub-literature on the differences, which Wight (1991) actually points out when introducing them. While realism is an understandable label, rationalism and revolutionism are not satisfactory either because they also do not match the contents of the ideas allegedly associated with them. Therefore, the distinction between international system, international society and world society will be used here. However, even these labels do not necessarily represent complete distinct traditions. They should be viewed as simple groupings of a few common threads of thoughts (of a number of theorists) over a long period of time. Let's see which tradition matches best with classical liberalism.

World society tradition

The world society tradition draws together ideas about the revolutionary replacement of the world of states by some form of world community. Buzan (2014) explains that Wight originally included a motley crew of anarchists, communists and liberals under this tradition. Yet, over the past few decades it has become synonymous with most of the liberal IR theories considered in chapter 10. World society is seen as the highest moral goal in international relations, with a quest for morality to replace power as the most important motive for action in world affairs. There is a strong belief in the harmony of interests between people. The emphasis is on replacing states with transnational and international organisations, or even federations, based on international law. Given this, it is clear that classical liberalism does not fit the world society tradition.

International system tradition

When comparing classical liberalism with the international system tradition, it should be remembered that this tradition is often equated with realism, since it regards international politics as a continuous struggle for power. Despite the existence of the United Nations and other international organisations, no authority is thought to exist above sovereign states. Therefore, states find themselves in an anarchical situation with a fundamental security dilemma. International politics is seen as a zero-sum struggle for power, i.e. a gain for one state leads to a loss for another since their interests are seen as mutually exclusive. Realists view human nature as selfish, with no real place for moral concerns in world politics or for international legal arrangements to steer the behaviour of states. While these arrangements do exist, states are seen to participate as long as it serves their national interests, but a *raison d'état* is usually used to justify ignoring international law and international organisations.

There are many strands to realism. The most fundamental divide is between classical realism – associated with Thucydides, Machiavelli, Hobbes, Jean-Jacques Rousseau and Max Weber – and neorealism, or as Waltz (2010) called it structural realism. Other thinkers associated with the international system tradition are modern writers such as E. H. Carr, George Kennan, Reinhold Niebuhr, Henry Kissinger and Hans Morgenthau. Realism is often associated with conservatism and in recent years with neoconservatism.

Neorealism emphasises the constraints that the international political system places on the actors. The units (states) do not differ functionally but have different capacities. Balance of power is the most rational policy for states to pursue. Often neorealism adopts a natural science methodology in IR analysis, such as game theory or the Prisoner's Dilemma, or the testing of hypotheses through the use of statistical methods and data sets. There is far less attention paid to economic issues or the possible influence of domestic actors on international relations.

Some readers may question the difference between classical liberalism and realism. After all, realism and the classical liberalism described thus far appear to share a number of ideas, such as the central place for the nation state in world politics, the appreciation of the balance of power and the recognition that war is sometimes inevitable. However, this is only part of the story. Classical liberals are more positive about the possibility of international order than international system thinkers. Their concern is with individual liberty rather than the interests of the state, and they are less eager to accept the principle of great power management, as all states should have their right to sovereignty respected. International law and organisations, while valuable, should be limited to protect individual liberty. They cannot just be discarded for reasons of state. World politics is not outright anarchy but an anarchical society of nation states with a place for moral concerns and rules such as keeping your promises.

Classical liberals embrace a realistic view of human nature, while realists have a more negative view. The difference is that classical liberals appreciate the social nature of humans and reject the idea that humans are inherently selfish. For classical liberals, the society of nations offers the most stable international order, which in turn secures individual liberty, or at least provides the best international conditions to secure liberty. Whether this will actually be the case for the individual of course depend on domestic arrangements. For classical liberals, world politics organised as an anarchic society of nations is preferable to an international system of outright anarchy, which is often less predictable and less open to classical liberal ideas, such as globalisation, free trade, war bound by just war rules, international law with some strength, functional international organisations, and the absence of government-to-government development aid, which is often doled out for power political reasons.

This also shows in the analysis of the work of Hume and Hayek. In an informative comparison of the ideas of Machiavelli and Hume, Whelan (2004) explains that Hume explicitly distanced himself from Machiavellian ideas such as the endorsement of conquest and expansion of national territory, the relation between military conquest and the greatness of the nation, wars that went beyond the need to keep the balance of power, or discarding the law of nations and the just war ideas. Whelan therefore concludes that 'Hume was not a forerunner of Hans Morgenthau's modern international realism'.

Hayek rejected the Hobbesian idea that the international sphere was a war of all against all. In *The Road to Serfdom* he also criticised the notable classical realist E. H. Carr as 'one of the totalitarians in our midst' in disagreeing with Carr's idea to subordinate morality in world politics. Hayek called for keeping your promises in international politics, or the principle of *pacta sunt servanda*. He also opposed Carr's idea that 'war was the most powerful instrument of social solidarity' and did not share his optimism about containing nationalism (Van de Haar 2009).

Classical liberalism as (pluralist) international society theory

Classical liberal ideas fit better with the international society tradition, which is seen as a middle position in English School theory. As indicated above, this tradition analyses international relations in terms of a society of states with common rules and institutions and a basis in natural law. Contrary to the international system tradition, it argues that states are not perpetually engaged in power struggle and war, and that some rules and morals apply. Contrary to world society tradition, it points out that sovereign states are and will remain the most important actors in international relations. Bull (1977) emphasised a number of institutions that foster order in world politics, including war, the balance of power, international law and diplomacy. He saw war as a necessary evil that helped shape international society while the balance of power brought about a form of international order. In addition, he also saw that the great powers had a special responsibility to maintain order, aided by international law and diplomacy. Bull divided the international society tradition into two camps. He labelled international society scholars who lean towards the international system as 'pluralists', while those who lean towards world society as 'solidarists'. The difference, as Buzan (2004, 2014) explains, is between the idea that 'order is in important ways a prior condition for justice' and the idea that 'order without justice is ultimately unsustainable'.

Mises and Hayek lived when Bull, Wight and the other first-generation English School members wrote about international relations theory. Yet there is no indication they ever met or were influenced by each other. However, there is clearly an agreement between classical liberals who wrote about international politics and the pluralist international society tradition. Both see the world as a society of states, with international conflict and war as a regrettable yet inevitable feature that needs to be bound by the rules of a just war. Both value balance of power, diplomacy and limited amounts of international law and organisation. There is also a direct appreciation of the Grotian notion of a just war.

Hume (1998) advocated just war, the international society of nations, and remarked that 'Hobbes's politics are fitted only to promote tyranny, and his ethics encourage licentiousness'. The state of nature was just 'philosophical fiction'. Hume's support for the balance of power was based on the insight that it fostered order which enabled rules and allowed nations to pay more attention to liberty, prosperity and benign modest social change through free trade. Smith also emphasised the value of international society, the need for just war and adherence to the principles of natural law (as in the work of Grotius) while acknowledging the role of the balance of power, the inevitability of war and the need for defence. Mises rejected all offensive war, endorsing the institutions of the international society of states, and was positive about Grotius's theory of war and peace. In general, he thought a commonwealth of free nations would constitute the best international order. Hayek (2007) famously wrote that his goal for international affairs was 'neither an omnipotent super state, nor a loose association of free nations, but a community of nations of free men'.

In conclusion, in the English School framework, classical liberalism belongs to the pluralist international society tradition. Pluralist, because classical liberals strongly believe order is a prerequisite for individual freedom. Other liberal IR theories make a better fit with the world society tradition. Also, it is just as important to conclude that classical liberalism is not some form of realism or makes a fit with the world system tradition.

Classical liberal foreign policy

Going from theory to practice is never straightforward, and the move from international relations theory to foreign policy is no exception. In addition, it should be noted that foreign policies are hardly ever designed and implemented from a blank slate. Any change in existing foreign policy commences with today's global situation, with individual nations having evolved their foreign policy in response to changes in the international political environment. Therefore, attempts to design a real classical liberal foreign policy on a purely theoretical basis are certain to fail. The best one can do is provide general guidelines or principles that can then be tailored to the specific circumstances of an individual state.

As will be shown in the next chapter, the impracticability of presenting an overall classical liberal foreign policy that is applicable to all countries means that it is a mistake to copy the recommendations of American libertarians and classical liberal think tanks from the American situation and apply them to other countries. Often their ideas do not fit the foreign policies of other countries. The US is in many ways an exception since it has tremendous military might, a truly global presence in diplomacy and international organisations, one of the largest economies and a massive cultural influence. Most other countries do not and probably will never be able to achieve this.

Nevertheless, it is also worth bearing in mind that foreign policies are not set in stone since the politics of foreign policy is always characterised by change (Hill 2003). This change can be the result of responses to events, such as the Russian war against Ukraine, which made Finland and Sweden conclude it was time to become NATO members, or responses to global changes, such as the end of the Cold War or climate change. There are also domestic factors that demand or foster changes in foreign policy. For example, a change of government, the role of other ministries, pressure groups, etc. Hence, the inclusion of classical liberal elements into the foreign policy of a particular country could be the result of change 'from within' and 'from outside'.

Some of the most important general guidelines for a classical liberal foreign policy, which are not meant to be exhaustive, are as follows:

- National and global security remain the pillars of any foreign policy, because of the security dilemma all countries face. Balance of power politics, including the active seeking and paying of one's dues in military alliance(s), is called for. International order is most important for individual liberty.
- War is sometimes inevitable as an instrument of foreign policy, but it should only be conducted according to just war principles. Military intervention is hardly ever justified and the chances of long-term success are slim anyway. This should not be confused with justified military operations meant to keep order in the globalised world, such as protecting the sea lanes against piracy.
- Diplomacy remains a useful tool in day-to-day international governmental contacts, as well as offering consular services and enabling travel and trade between nations. There can be debate over whether diplomacy should include trade promotion

initiatives and whether it should protect existing trade relations. After all, managed trade or trade policy is mercantilist in nature.

- The protection of citizens and border protection are core tasks. Once again, it should be noted that classical liberals have different views about the openness of borders.

- The reach, content and number of international laws and international organisations should be minimised since they can be a threat to individual liberty. Functional agreements and regimes are, of course, possible and desirable. There will also be many 'grey areas', but the international treaties should always have a clause for countries to leave.

- Further promotion of free trade should be a prime objective. Ideally, this would be trade without any government intervention. The WTO is a second-best, suboptimal option, yet it provides a dispute settlement mechanism and its rules and agreements apply to all members. Bilateral trade treaties are acceptable, but the goal should be to integrate them into multilateral agreements via the WTO or to get rid of them altogether in favour of real free trade.

- Classical liberals are sceptical of government-to-government development aid, and believe that the World Bank and other governmental development aid organisations should be abolished. Emergency relief is the exception. Voluntary donations and NGOs are acceptable. Classical liberals oppose government funding of NGOs.

- The United Nations, especially the Security Council, should – at the very least – be reformed to take account of the power relations in the modern world. Of course, in Britain and France, some classical liberals may feel a patriotic duty to support their governments' continued status as permanent members, but in the big global picture, this is anachronistic. The UN is valuable: it provides a place where countries can meet and discuss current affairs, and work to prevent conflicts from unnecessarily increasing in intensity. However, many of its daughter organisations could either be abolished or trimmed to their core, so that they remain strictly functional. The UN Human Rights Council has become an outright anomaly with its membership of autocracies and dictatorships.
- Regional cooperation is desirable as long as it complies with the general classical liberal guidelines of limited state interference in the life of individuals.

12 LIBERTARIANS AND IR

It is hard to find any presentation of libertarian ideas on international relations in IR theory textbooks, journals or other publications. While there are libertarians who write about international relations, most of their ideas diverge much more from mainstream IR theory than of those libertarians who write about economics or philosophy. In contrast to classical liberal ideas about international relations, many libertarians tend to focus on a few selected topics relating to foreign affairs. Murray Rothbard, one of the best known anarcho-capitalists, was therefore right in his remark that international relations, or questions of war and peace, were all too often ignored in libertarian thought (Rothbard 2000).

This chapter examines some of the main libertarian ideas on international politics, including their historical roots, differences with classical liberalism, application to some specific issues as well as the main ideas of Ayn Rand. Since this is a brief overview no claim to completeness can be made.

Like classical liberalism, libertarianism is a very broad church. In fact, in modern American parlance classical liberalism and libertarianism are sometimes considered synonyms for each other, while individuals also use the labels interchangeably, which makes it even more difficult to distinguish between the two (see Van de Haar 2015). As was discussed in chapter 2, in this book libertarians are mainly seen as anarcho-capitalist and minarchist thinkers (believers in a minimal state). Although it remains difficult to pinpoint and categorise the ideas of those writers who identify as libertarians the portrayal in this chapter claims to offer a fair overview.

Historical traces

Nineteenth-century thinkers such as Cobden and Spencer are sometimes considered part of the libertarian spectrum with a strong cosmopolitan outlook. However, some of their ideas also serve as inspiration for classical liberals.

Richard Cobden (1804–65) is quintessential in this sense. Together with John Bright and the Anti–Corn Law League, his pleas for free trade inspire both classical liberals and libertarians. Others of his ideas are easier to relate to libertarianism than to classical liberalism, mainly due to his opposition to the role of the state in international relations. Cobden is also largely responsible for the early misrepresentation of Adam Smith's ideas on international affairs. He presented himself as a 'Smithian' calling for the founding of Adam Smith Societies to honour the great Scot. However, he erroneously claimed that Smith believed that free trade would lead to peace, was leaning towards pacifism, considered that a harmony between states was possible, and that military expenditure should be cut (Van de Haar 2010).

This idea that free trade promotes peace through developing interdependencies between countries was central to members of the Anti–Corn Law League, many of whom were also convinced that trade would spread civilisation and Christianity. There were close ties between the League and the pacifist movement. Cobden never called himself a pacifist but remained involved in peace conferences and often spoke about the goal of international peace. Cobden was an international celebrity, travelled widely, became a Member of Parliament, and negotiated the Cobden–Chevalier free trade treaty with France in 1860. International affairs of higher moral standards remained an important idea for Cobden. He advocated non-intervention, including in the colonies of other states, and international arbitration to settle disputes. Cobden opposed a foreign policy based on the balance of power, the procurement of armaments, empire and colonisation. He contended that war served the elites, not the middle classes (also Hammarlund 2005).

Most of these ideas were embraced by British social liberals and socialists, from the last quarter of the nineteenth century onwards, and even American president Woodrow Wilson was 'a self-confessed Cobdenite'. However, while Cobden was not a prominent figure in the twentieth century, his ideas still fuelled some libertarian viewpoints on international relations. According to Cobden scholar Frank Trentmann (2006), this was mainly due to the references Ludwig von Mises made to Cobden and 'Manchesterism', although Mises actually disagreed with many of Cobden's ideas (Van de Haar 2009).

Herbert Spencer (1820–1903) was also very influential, both in the UK and the US. His work can arguably be described as proto-libertarian. One of his ideas that was influential for the development of libertarian thought was that there is nothing eternal about a state or a certain form of government. Individuals are born into a certain society without making a conscious choice. According to Spencer, every individual has the right to withdraw from the state's authority as soon as this state offers no, or insufficient, protection in exchange for the taxes it levies (Spencer 1982). This idea of the right to secession is popular among many libertarians.

In the context of international relations, Spencer called for completely free trade, war for strict self-defence only and the abolition of empire. He saw empire as 'political burglary', which did not bring any economic benefits and fostered a bad, militaristic mentality in the mother country. Although war was often part of the early stages of the development of societies, in modern societies it was the reverse of individualism, offering the state opportunities to increase its authority and control over many aspects of individuals' lives. Spencer undertook a life-long campaign to end warfare, initially urging England to lead the pacifist trail. He saw international arbitration as a much better way to solve international conflicts. In 1881, Spencer launched the Anti-Aggression Association to promote these ideas and mobilise what he believed would be 'the large amount of anti-war feelings, especially among the artisan classes and the great body of dissenters'. Late in life, Spencer was less optimistic about the chance to rid humanity of war. Yet he opposed the Boer War and urged Andrew Carnegie, the pacifist Scottish–American industrialist and philanthropist, to become involved and spend money on an anti-war campaign (Francis 2007).

Classical liberal versus libertarian IR

Cobden's and Spencer's ideas have influenced libertarian ideas on international relations, especially scholars in US think tanks including the Cato Institute, the Mises Institute, the Independent Institute, The Future of Freedom Foundation and the Foundation for Economic Education. These are the think tanks that have produced the most libertarian output on international relations in the past decades. Yet libertarian ideas have been less prevalent in academia. An exception is Lomasky and Tesón's *Justice at a Distance: Extending Freedom Globally* (2015), a rare academic book on a libertarian view of international relations.

While there are debates among American libertarian international relations writers, US think-tankers have developed a cosmopolitan foreign policy outlook based on a mixture of libertarian principles, the writings of Cobden and Spencer and some of the classical liberal ideas presented in this book. Such writers often also refer to American history, especially the ideas of limited government popular among some of the Founding Fathers. An example of this is Preble's *Peace, War, and Liberty* (2019), which was written when the author was still at the Cato Institute.

These authors tend to be more focused on US foreign policy and less concerned with, and maybe even largely unaware of, non-US classical liberal views on international affairs. This makes it sometimes difficult for others to understand classical liberal and libertarian ideas on international relations and to distinguish between them.

Of course, there are a number of ideas classical liberals and libertarians agree on, such as the expansion of free trade, globalisation as a force for good and criticism of international governmental organisations such as the IMF, UN, World Bank or WTO. However, some classical liberals do support the work of the WTO as a practical, suboptimal way to increase free trade. Libertarians and many classical liberals also agree that non-intervention should be a goal in foreign policy, although libertarians tend to be more absolute on this issue. They are also both opposed to government-to-government development aid, and acknowledge the large costs of war in terms of lives lost, economic loss, devastation and increased power by the state over the lives of individuals. Classical liberals also often agree with libertarian critiques of specific US foreign policies.

Yet classical liberals feel many libertarians fail to recognise the natural and emotional bond between individuals and the nation, which ensures that states remain important actors in international relations. While libertarians recognise that humans are not angels, classical liberals are not convinced that libertarians really acknowledge the consequences of human nature for world politics. Classical liberals tend to be less supportive of secession from state activities and sometimes view libertarian thought on international relations as purely isolationist, founded on the erroneous belief that if one group of people unilaterally agrees not to interfere with others, all will be fine.

The American roots and audience of most libertarian writers show clearly in their neglect of the question: 'What if non-alignment leads to others aligning against you?' It is easy to assume your defence will be able to deal with all possible threats and violence if you are a citizen of the US, the world's foremost power. However, for most other countries, international alliances are seen as an essential feature of foreign policy.

Libertarians are most often proponents of open immigration and strongly believe in the relationship between trade and peace. It is interesting that many libertarians share the views of many social liberals on issues such as open immigration and pacifism, as well as opposition to offensive war, the balance of power and the central role of nation states in world politics. They are also suspicious of the role of powerful elites in foreign policy making.

War

Considerations of war play an important role in libertarian thought about international relations. Some scholars think war can be eradicated if all states pursue isolationist policies, while others are less optimistic. Just like classical liberals, Rothbard (2002) explicitly embraces the just war tradition of the Spanish Scholastics and Grotius. War cannot be abolished but instead should be constrained by 'limitations imposed by civilisation', in particular a prohibition on targeting civilians in war and the preservation of the rights of neutral states. Just wars are those where people defend themselves against an external threat of coercive domination, or try to overthrow foreign domination. An unjust war is one that seeks domination over other people or attempts to retain existing coercive authority.

In the libertarian theory of war, the central idea is the application of the libertarian axiom that no one may threaten or commit violence against another person or their property. Only if that principle has been violated is direct action against the offender warranted. The rights of innocent people should not be violated in any retaliatory actions. Hence, for many libertarians the use of weapons of mass destruction, which do not discriminate between belligerents and innocents, is a criminal act.

For Rothbard, the principle of neutrality is of great importance since it enables states to stay out of conflicts. He views neutrality as an act of great statesmanship but regrets that this is no longer generally recognised due to governments believing that collective security arrangements are needed, or that there is a moral obligation to impose democracy or human rights throughout the globe. He believed that 'rights may be universal, but their enforcement must be local' by the people who feel their rights are infringed upon (Rothbard 2003). In a world of states, Rothbard felt that libertarians should try to pressure the government to absolutely avoid war and to disarm 'down to police levels'. When in conflict, governments should be pressed to negotiate a peace, immediately declare a ceasefire, keep non-combatants out of the fighting, or stay neutral if not directly involved. These guidelines, which Rothbard did not mind being labelled as 'political isolationism', would enable peaceful international coexistence. However, other libertarians, such as John Denson, argue for 'well-armed neutrality' (Denson 2003).

In an argument unique to libertarians, thinkers such as Stromberg (in Hoppe 2003b) emphasise that militias employing guerrilla tactics are the best form of national defence. They point to the alleged successes of various guerrilla groups in fighting state armies. They view guerrilla tactics by militias as a way to defend states and territories in a stateless world or as a means of enabling territories to secede from a larger territory. Militias are flexible, can be privately financed and are difficult to beat. As expected, this analysis of the efficacy of militias has been challenged. For example, militias often take cover behind civilian populations, which violates a key principle of a just war. Also, militias have been known to seek to expand their activities beyond the territories they initially defend and to capture civilian populations that do not support their activities, or to engage in acts of aggression against other militias.

Libertarians reject the idea that a war economy needs government command and oppose other controls during times of war, such as state propaganda and restrictions on free speech. For example, Higgs (2005) believes that there are always economic costs involved in warfare and that there is no such thing as 'war prosperity'. Governments use the public demand for their services in times of crisis and never fully restore the pre-crisis situation, in the process expanding their powers and state influence over individuals. In this sense, war is the ultimate crisis, and the war against terrorism is no exception. Higgs as well as fellow libertarians Ebeling and Hornberger (2003) have pointed out that, since the 9/11 attacks, there has been an enormous expansion of government expenditure and a limitation of civil liberties, especially in the US.

Private defence

Hoppe (2003a) defines government as a 'compulsory territorial monopolist of protection and jurisdiction, equipped with the power to tax without unanimous consent', and believes that it will fail in the provision of defence. It is just another monopolist that charges a high price, while delivering poor-quality service. Both Hoppe (2003a) and Block (2003) attempt to show that defence is not a public good and that private defence may be a better alternative. This is not a new idea and builds on the writings of nineteenth-century economist De Molinari (1849), who argued that no state is needed to provide military security against attacks.

Hoppe (2003a) believes that private insurance companies could be employed for external defence. In the case of an attack by a foreign state, the territory would be defended by a combination of an armed citizenry and an alliance of insurance and reinsurance companies. Their efforts will by definition be superior to those of the aggressor state, because by definition a state would be inefficiently organised compared to private organisations. Hoppe asserts that the insurance companies would be aware of all possible risks beforehand, and that these risks would be taken into account in the calculation of the insurance premium to be charged. He expects insurance companies to counter aggression and also 'possibly incite the liberation and transformation of the state territory' of any offending state. He believes that a world with private security companies would be less war prone since it is in the interest of security companies to avoid fighting.

While Hoppe (2003a) makes a number of assumptions that are challengeable, it should also be noted that his approach appears to be more an exercise in economics than in international relations theory. While there are numerous insurance companies dealing with political risk and uncertainties, as well as private security companies employed in conflicts, most conflicts are still between states, or between states and guerrilla or terrorist organisations looking for a part, or a change in the state-controlled resources or policies. Hoppe's libertarian economic view of defence also overlooks differences in defence capabilities between territories due to manpower, geography, geology or economic strength. These factors can be serious constraints on group action, whether financed privately or by taxpayers. It also does not take into account other characteristics at the international level, such as the balance of power. The whole idea is also based on a strictly rational view of human nature that is questionable, to say the least. And it assumes, without any evidence, that a defence based on private insurance would be superior to a state-financed military apparatus, often with deep pockets.

US foreign policy

Many libertarians focus on US foreign policy in their writings on international affairs, arguing that military alliances drag the US into unnecessary wars that do not constitute a clear and present danger to its security (see, for example, Higgs 2005). They believe that the US should not attempt to be 'the policeman of the world' and that the prime and perhaps only objective of US defence should be the protection of US citizens and their property from attack by hostile states. The US should also not maintain a 'global empire' through military bases scattered around the world.

In addition to the costs in terms of lives and money, they view most if not all US foreign intervention as contrary to the intentions of the Founding Fathers. Libertarians such as Carpenter (1989, 2002) and Eland (2004) argue that the credibility of US support for democracy is undermined when the US supports dictators remaining in power, which in turn leads to anti-Americanism and even terrorist attacks against America or American citizens. They believe that the US should withdraw from alliances such as NATO and the defence treaties with Japan and South Korea. Carpenter (1995), in a more belligerent analysis than most libertarians espouse, asserts that America must become the 'balancer of last resort in the international system.' This would mean that the US retains enough military force to stop 'unusually potent expansionist threats' to international stability, that cannot be contained by alliances with other, smaller powers. However, this could also be seen as a call for US foreign intervention in certain circumstances.

Other libertarians, such as Rockwell (2003) and Ebeling and Hornberger (1996), raise concerns about the continued existence and expansion of American power and often argue in favour of non-interventionism. They believe that the desire to keep 'an empire' is a major cause of mistakes in US foreign policy. They often judge the size of the alleged American empire by the number of US military posts abroad or the number of foreign interventions. They also appear to assume that all US action is against the will of the people concerned. For example, they view the US military as occupying Japan due to the large military base in Okinawa, even though the Japanese government and many Japanese citizens might view the US military presence as security against an increasingly more powerful Chinese military. However, as indicated, it should be noted that some classical liberals would also question the necessity of US foreign intervention.

Ayn Rand[2]

Ayn Rand became one of the best-known libertarian thinkers through her novels and non-fiction writing. Despite the obvious minarchist traits of her work, she did not want to be grouped with others and defined her own philosophy as Objectivism. Her opinions and ideas on international relations differ substantially from other libertarians (Van

2   This section is based on Van de Haar (2019).

de Haar 2019). Foremost, Objectivism is a moral theory, but politics does play an important part. World politics was one of Rand's main political concerns. As a Russian *émigré* she deeply hated communism as much as, if not more than, other collectivist theories, such as fascism, national socialism, but also nationalism or ethnicism.

Her Objectivism viewed defence against foreign invasion as one of the three justified government functions (the others being the police and judiciary). Objectivists believe that the government should hold the monopoly on retaliatory violence. Rand never endorsed private ownership of weapons or privately funded defence. She also saw national culture and the subconscious 'sense of life' as important for individuals. Rand wanted to reinvigorate the 'individualist spirit' among Americans, which she felt was under threat, and saw sovereignty as a right that can be earned or forfeited. It can be earned and morally secured if a nation fully respects the principle of individual rights. In that case other nations have to respect its sovereignty. However, if a country violates the rights of its citizens, it loses its right to sovereignty. In characteristically strong words: 'a nation ruled by force is not a nation but a horde, whether led by Attila, Genghis Khan, Hitler, Khrushchev, or Castro' (Rand 1964). She believed that dictatorships are outlaws that can be invaded at will as long as the invading state has the intention of restoring individual rights. Therefore, the right to self-determination is only a right to become a free nation.

Rand also despised realists such as Kissinger for what she saw as a disregard for morality in international politics or their support for dictatorships, out of 'practical' or 'strategic' thinking. She opposed President Nixon's change of US foreign policy towards China, which she saw as a betrayal of Taiwan. Her language was very belligerent, especially in the Cold War fight against communism or against nations that disregarded individual liberty, including China. She did not expect a peaceful world, because only fully rational people could achieve that, and she viewed the world as full of irrational behaviour and grounds for conflict. She believed that war is part of human nature but should only be used in retaliation and against the party that initiated violence. An Objectivist principle is that no person or state has the right to initiate violence. Therefore, international order depends on a strong defence against evil forces. In this context Rand valued the balance of power in politics.

Rand was dismissive of the pacifying effects of liberal policies or theories, viewing the nuclear arms race as necessary, pacifism as evil, and seeing no need for the reduction of armed forces. She explicitly criticised liberal internationalism, arguing against the notion that a nation's sovereignty and interest should be sacrificed for world community, which Rand saw as contrary to the rights and interests of individuals. She also opposed the United Nations, not least because it provided the Soviet Union and other dictatorships with prestige, which she saw as undeserved.

Like most liberals, Rand opposed the military draft and objected to the Vietnam war. This made her temporarily popular among activists in the 1960s and 1970s, who usually opposed her views in favour of free trade and capitalism and against development aid and international law. However, she also believed that once treaties were signed, they should be adhered to. To conclude, her viewpoints are best described as a mix of classical liberal and some libertarian ideas, with belligerent, undiplomatic views on certain topical issues in world politics, which distanced her from other libertarians.

Libertarian international relations vs. classical liberal international relations

While there are overlapping views between classical liberals and libertarians, not least since some ideas are derived from the same thinkers, most libertarian views on foreign policy originate from the US, which gives American libertarians a US-centric view of the world. As we have seen, many libertarians view foreign policy in economic terms and oppose the costs of war and military intervention and are often concerned about a self-serving 'American empire'. The most fundamental difference is the view of human nature and the assessment of the role of reason in predicting and explaining human behaviour.

13 CONCLUSION

This book examined the contours of the classical liberal theory of international relations. This theory is not based on reasoning from abstract principles, but instead is mainly 'distilled' from the writings of four classical liberal authors: David Hume, Adam Smith, Ludwig von Mises and F. A. Hayek. While the first two wrote about international relations in the eighteenth century, and the latter two wrote in the last century, this book has attempted to show their relevance to international relations today. In doing so, a theory from first principles is presented, starting with the individual and the groups most near to her before expanding to the international level. In this respect, there is no difference between domestic classical liberalism and international classical liberalism: securing and fostering individual liberty is the main aim.

The classical liberal theory of international relations presented here is based on the examination of individual human nature and the relationship between the individual and groups, particularly the nation and the state. It is natural for classical liberals to regard the nation state as the main actor in international relations, mainly because they assume, in most people, an emotional and psychological bond between the individual and her nation. On the basis of its realistic view of human nature, classical liberalism also recognises the inevitability of conflict and war and aims to deal with their occurrences, however regrettable. Classical liberals believe that international order is achieved through many actions and sources. There is no single magic recipe. Therefore, you need spontaneous ordering forces, such as the balance of power, but also limited and strong international written rules. International law and a limited number of functional international governmental organisations may be needed to protect classical human rights and agree on (mostly) functional issues arising in a world of states. Classical liberals are also in favour of global free trade and globalisation, which bring many benefits, including lifting many people across the world out of poverty. However, free trade should not be seen as inherently fostering peace. Classical liberals tend to be sceptical of government-to-government taxpayer-funded international aid, but they may be in favour of aid for short-term disaster relief out of humanitarianism. In line with their belief in voluntary action, they do not object to individuals supporting non-state charities, or NGOs. Lastly, classical liberals believe in restraint. Military intervention should be an exception. There is also no place for imperialism, let alone collectivist notions such as nationalism. The issue of immigration is still open for debate, but in the classical liberal theory presented here, limits on immigration are completely acceptable.

An application of Freeden's morphological analysis (1996) finds that in domestic politics classical liberalism differs substantially from social liberalism, libertarianism and conservatism (Van de Haar 2015). This difference also applies to international politics, where the analysis also reveals a substantial difference between classical liberalism and other existing liberal IR theories. These existing liberal IR theories, as well as those taught on academic IR courses, tend to focus on the mitigation of alleged negative effects of the centrality of nation states in global politics, including power politics and the occurrence of war. On the basis of an alleged global harmony of interests between people, these liberal IR theories introduce a number of far-reaching proposals regarding international and supranational law and international governmental organisations. Some even favour world society or world (federal) government. In general, these liberal internationalists argue that replacing power politics with morality should be the central idea in international politics. Classical liberals shy away from such ideas, which they see as idealistic and unrealistic. They accept humans and the world as they are, instead of how we would like them to be.

The differences between classical liberals and libertarians are just as pronounced. Libertarians tend to be more sceptical of engaging in wars or any military intervention, no matter how just the cause may appear. They believe that such neutrality, or even complete isolationism, creates a more peaceful world. However, it must be noted that most libertarian writers are Americans, writing mostly for American audiences. Given that the US is the world's strongest economic, military and arguably cultural power, this will obviously shape US libertarians' views, while at the same time limiting the applicability of their ideas to the rest of the world. Some libertarians also write about international affairs from an overriding economic transactional perspective, which focuses on issues such as the private production of security, but overlooks other important factors such as the multiple causes of war.

We have also briefly discussed the even bigger difference between classical liberalism and non-liberal IR theories, and compared classical liberal ideas to the three traditions of the English School of international relations. Classical liberalism and the (pluralist) international society tradition make the best fit, while most other liberal IR theories are better seen as ideas that fit the world society tradition.

Realism, one of the main IR theories, is an international system theory. It is true that realism and classical liberal IR theories share some common ideas, such as the central role of the nation state in world politics, the appreciation of the balance of power and the recognition that war is sometimes inevitable. However, the classical liberal view of international relations should not be seen as simply a variant of realism.

Classical liberals are more positive about the possibility of international order than international system thinkers. Their concern is with individual liberty rather than the interests of the state, and they are less eager to accept the principle of great power management since many classical liberals believe that nation states should have their right to sovereignty respected. Classical liberals also believe that some international law and international (governmental) organisations are valuable but should be limited to protect individual liberty. Classical liberals do not view global politics in terms of outright anarchy but as an anarchical society of nation states where there is room for moral concerns.

Most fundamentally, classical liberals embrace a realistic view of human nature, in contrast to the negative view of the realists. The differences are the classical liberal appreciation of the social nature of humans and the rejection of the idea that humans are inherently selfish. For classical liberals the society of nations offers the most stable international order, which is the best way to secure, or at least provide, the best international conditions to achieve individual liberty. This also depends on domestic arrangements. In terms of domestic politics, realism is often associated with conservatism or neoconservatism.

International politics influences the lives of many people in this world. Classical liberalism is a universalist theory, claiming that its ideas are applicable across the globe. This book has argued that classical liberalism has a distinct view of international affairs and deserves its own place in liberal international relations theory. However, it should be acknowledged that much more work needs to be done. Surely, not all people who call themselves classical liberals will agree on issues such as the European Union or immigration. There need to be more analysis, research, discussions and debates among liberals of all persuasions, in academia and in other forums. More classical liberal authors need to be studied to understand and incorporate their ideas on international relations. Many of the building blocks discussed in part II also require further study and consideration. This book made a cautious attempt to apply classical liberal ideas to contemporary foreign policy discussions. More needs to be done on connecting classical liberal ideas to current international affairs in order to strengthen the appeal of classical liberalism to a wider audience.

It must be emphasised though that classical liberalism is not just a domestic theory with very little to say about international affairs. Classical liberal writers have developed unique ideas on world politics that can be applied to contemporary foreign affairs issues. Compared to other liberal theories of international relations, classical liberalism is as important and relevant. There is far more to liberalism in IR than hitherto thought.

LITERATURE AND FURTHER READING

Please note, most chapters use parts of:

Van de Haar, E. R. (2009) *Classical Liberalism and International Relations Theory: Hume, Smith, Mises, and Hayek.* New York and Basingstoke: Palgrave Macmillan.

and

Van de Haar, E. R. (2015) *Degrees of Freedom: Liberal Political Philosophy and Ideology.* New York and London: Routledge.

Please refer to these books for further discussion and detailed references. Please also be aware that there are great online sources, where many full books are found, often free: e.g. the Online Library of Liberty, Adam Smith Works, IEA.org, Mises.org, Liberty Fund.

Chapter 1   Introduction

Brown, C. and Eckersley, R. (2018) *The Oxford Handbook of International Political Theory.* Oxford University Press.

Cunliffe, P. (2020) *The New Twenty Years' Crisis: A Critique of International Relations 1999–2019.* Montreal and Kingston: McGill-Queen University Press.

Griffiths, M. (2011) *Rethinking International Relations Theory.* New York and Basingstoke: Palgrave Macmillan.

Halliday, F. (1994) *Rethinking International Relations.* Basingstoke and London: Macmillan Press.

Jönsson, C. (2018) Classical liberal internationalism. In *International Organization and Global Governance* (ed. T. G. Weiss and R. Wilkinson), 2nd edn, pp. 109–22. Routledge.

Chapter 2   Liberalisms (and conservatism)

The contours and some content of this chapter are based on Edwin van de Haar, The meaning of 'liberalism', 25 April 2015 (www.libertarianism.org).

Ashford, N. (2003) *Principles for a Free Society.* Stockholm: Jarl Hjalmarson Foundation.

Berlin, I. (1969) *Four Essays on Liberty.* Oxford University Press.

Butler, E. (2015) *Classical Liberalism: A Primer.* London: Institute of Economic Affairs.

Butler, E. (2019) *School of Thought: 101 Great Liberal Thinkers.* London: Institute of Economic Affairs.

Freeden, M. (1996) *Ideologies and Political Theory: A Conceptual Approach.* Oxford: Clarendon Press.

Freeden, M. (2003) *Ideology: A Very Short Introduction.* Oxford University Press.

Kirk, R. (1985) *The Conservative Mind: From Burke to Elliot.* Washington: Regnery Publishing.

Mises, L. von (1996 [1949]) *Human Action: A Treatise on Economics.* San Francisco: Fox & Wilkes.

Nisbet, R. (1986) *Conservatism: Dream and Reality.* Milton Keynes: Open University.

Oakeshott, M. (1962) *Rationalism in Politics and Other Essays.* New York: Basic Books.

Scruton, R. (2001) *The Meaning of Conservatism.* Basingstoke: Palgrave.

Scruton, R. (2017) *Transcript: Point of View.* BBC Radio 4, 27 August.

Chapter 3   Scottish Enlightenment: David Hume and Adam Smith

Berry, C. J. (1997) *Social Theory of the Scottish Enlightenment.* Edinburgh University Press.

Fleischacker, S. (2004) *On Adam Smith's Wealth of Nations: A Philosophical Companion.* Princeton University Press.

Glossop, R. J. (1984) Hume and the future of the society of nations. *Hume Studies* X: 46–58.

Grotius, H. (2005) *The Rights of War and Peace* (3 books) (ed. R. Tuck). Indianapolis: Liberty Fund.

Haakonssen, K. (1981) *The Science of a Legislator. The Natural Jurisprudence of David Hume and Adam Smith.* Cambridge University Press.

Haakonssen, K. (2006) Introduction. The coherence of Smith's thought. In *The Cambridge Companion to Adam Smith* (ed. K. Haakonssen), pp. 1–21. Cambridge University Press.

Hardin, R. (2007) *David Hume: Moral and Political Theorist.* Oxford University Press.

Harris, J. A. (2015) *Hume: An Intellectual Biography.* Cambridge University Press.

Hume, D. (1932 [1727–76]) *The Letters of David Hume* (two vols) (ed. J. Y. T. Greig). Oxford University Press.

Hume, D. (1985 [1777]) *Essays: Moral, Political and Literary* (ed. E. F. Miller). Indianapolis: Liberty Fund.

Hume, D. (1998 [1751]) *An Enquiry Concerning the Principles of Morals* (ed. T. L. Beauchamp). Oxford University Press.

Hume, D. (2000 [1739]) *A Treatise on Human Nature* (ed. D. F. Norton and M. J. Norton). Oxford University Press.

Manzer, R. A. (1996) The promise of peace? Hume and Smith on the effects of commerce on war and peace. *Hume Studies* XXII: 369–82.

Mises, L. von (1996) *Human Action: A Treatise on Economics*. San Francisco: Fox & Wilkes.

Mossner, E. C. (1980) *The Life of David Hume*. Oxford: Clarendon Press.

Ross, I. S. (2010) *The Life of Adam Smith*, 2nd edn. Oxford University Press.

Smith, A. (1981 [1776]) *An Inquiry into the Nature and Causes of the Wealth of Nations* (ed. E. C. Mossner and I. S. Ross). Indianapolis: Liberty Fund.

Smith, A. (1982 [1759]) *The Theory of Moral Sentiments* (ed. D. D. Raphael and A. L. Macfie). Indianapolis: Liberty Fund.

Smith, C. (2006) *Adam Smith's Political Philosophy: The Invisible Hand and Spontaneous Order*. London and New York: Routledge.

Stevens, D. (1987) Smith's thoughts on the state of the contest with America, February 1778. In *Correspondence of Adam Smith* (ed. E. C. Mossner and I. S. Ross), pp. 377–80. Indianapolis: Liberty Fund.

Van de Haar, E. R. (2008) David Hume and international political theory: a reappraisal. *Review of International Studies* 34: 225–42.

Van de Haar, E. R. (2013a) Adam Smith on empire and international relations. In *The Oxford Handbook of Adam Smith* (ed. C. J. Berry, M. P. Paganelli and C. Smith), 417–39. Oxford University Press.

Van de Haar, E.R. (2013b) David Hume and Adam Smith on international ethics and humanitarian intervention. In *Just and Unjust Military Intervention: European Thinkers from Vitoria to Mill* (ed. S. Recchia and J. M. Welsh), pp. 154–75. Cambridge University Press.

Whelan, F. G. (2004) *Hume and Machiavelli: Political Realism and Liberal Thought.* Lanham: Lexington Books.

Chapter 4   Austrian School: Ludwig von Mises and F. A. Hayek

Boettke, P. J. (2019) *F. A. Hayek: Economics, Political Economy and Social Philosophy.* Basingstoke and New York: Palgrave Macmillan.

Butler, E. (1985) *Hayek: His Contribution to the Political and Economic Thought of Our Time.* New York: Universe Books.

Butler, E. (1988) *Ludwig von Mises: Fountainhead of the Modern Microeconomics Revolution.* Aldershot and Brookfield: Gower.

Caldwell, B. (2004) *Hayek's Challenge: An Intellectual Biography of F. A. Hayek.* University of Chicago Press.

Caldwell, B. and Klausinger, H. (2022) *Hayek: A Life, 1899–1950.* University of Chicago Press.

Dekker, E. (2016) *The Viennese Students of Civilization: The Meaning and Context of Austrian Economics Reconsidered.* Cambridge University Press.

Ebenstein, A. (2001) *Friedrich Hayek: A Biography.* New York and Houndmills: Palgrave Macmillan.

Ebenstein, A. (2003) *Hayek's Journey: The Mind of Friedrich Hayek*. New York and Houndmills: Palgrave Macmillan.

Feser, E. (2006) *The Cambridge Companion to Hayek*. Cambridge University Press.

Hayek, F. A. (1948) *Individualism and Economic Order*. University of Chicago Press.

Hayek, F. A. (1990) *Denationalisation of Money: The Argument Refined*, 3rd edn. London: Institute of Economic Affairs.

Hayek, F. A. (1994) *Hayek on Hayek: An Autobiographical Dialogue* (ed. S. Kresge and L. Wenar). University of Chicago Press.

Hayek, F. A. (1997) *Socialism and War: Essays, Documents, Reviews. The Collected Works of F. A. Hayek*, vol. X. University of Chicago Press.

Hayek, F. A. (2007) *The Road to Serfdom. Text and Documents. The Definitive Edition. The Collected Works of F. A. Hayek*, vol. II. University of Chicago Press.

Hayek, F. A. (2021) *Law, Legislation and Liberty. A New Statement of the Liberal Principles of Justice and Political Economy* (ed. J. Shearmur). *The Collected Works of F. A. Hayek*, vol. XIX. University of Chicago Press.

Hayek, F. A. (2022) *Essays on Liberalism and the Economy* (ed. P. Lewis). *The Collected Works of F. A. Hayek*, vol. XVIII. University of Chicago Press.

Hülsmann, J. G. (2007) *Mises: The Last Knight of Liberalism*. Auburn: Ludwig von Mises Institute.

Kirzner, I. M. (2001) *Ludwig von Mises: The Man and His Economics*. Wilmington: ISI Books.

Mises, L. von (1983) *Nation, State, and Economy: Contributions to the Politics and History of Our Time*. New York and London: Institute for Humane Studies & New York University Press.

Mises, L. von (1985) *Omnipotent Government: The Rise of the Total State and Total War.* Grove City: Libertarian Press.

Mises, L. von (1996) *Human Action: A Treatise on Economics.* San Francisco: Fox & Wilkes.

Mises, L. von (2000) An Eastern Democratic Union: a proposal for the establishment of a durable peace in Eastern Europe. In *Selected writings of Ludwig von Mises* (ed. R. Ebeling). Indianapolis: Liberty Fund.

Simon, J. (1996) *The Ultimate Resource 2.* Princeton University Press.

Van de Haar, E. R. (2022) Ludwig von Mises and Friedrich Hayek: federation as last resort. *Cosmos + Taxis* (10) 11+12, 104–18.

Chapter 5   Individuals: Human nature, natural and human rights

Berry, C. J. (1986) *Human Nature.* Basingstoke: Macmillan.

Coker, C. (2014) *Can War Be Eliminated?* Cambridge: Polity.

Donelan, M. (2007) *Honor in Foreign Policy: A History and Discussion.* New York and Basingstoke: Palgrave Macmillan.

Garrett, A. (2003) Anthropology: the 'Original' of human nature. In *The Cambridge Companion to the Scottish Enlightenment* (ed. A. Broadie), pp. 79–93. Cambridge University Press.

Hayek, F. A. (1993) *The Constitution of Liberty.* London: Routledge.

Hume, D. (1998) *An Enquiry Concerning the Principles of Morals.* Oxford University Press.

Hume, D. (1999) *An Enquiry Concerning Human Understanding.* Oxford University Press.

Hume, D. (2000) *A Treatise of Human Nature.* Oxford University Press.

Jackson, R. H. (2000) *The Global Covenant: Human Conduct in a World of States.* Oxford University Press.

MacMillan, M. (2020) *War: How Conflict Shaped Us.* London: Profile Books.

Mill, J. S. (1989) *On Liberty and Other Writings.* Cambridge University Press.

Mises, L. von (1985 [1927]) *Liberalism: The Classical Tradition.* Irvington-on-Hudson: The Foundation for Economic Education.

Mises, L. von (1996) *Human Action: A Treatise on Economics.* San Francisco: Fox & Wilkes.

Pinker, S. (2002) *The Blank Slate: The Modern Denial of Human Nature.* New York: Penguin Books.

Pinker, S. (2011) *The Better Angels of Our Nature: A History of Violence and Humanity.* London: Penguin.

Rosen, S. P. (2005) *War and Human Nature.* Princeton University Press.

Rubin, P. H. (2002) *Darwinian Politics: The Evolutionary Origin of Freedom.* New Brunswick and London: Rutgers University Press.

Smith, A. (1982) *The Theory of Moral Sentiments.* Indianapolis: Liberty Fund.

Thayer, B. A. (2004) *Darwin and International Relations: On the Evolutionary Origins of War and Ethnic Conflict.* Lexington: University Press of Kentucky.

United Nations (2021) Responsibility to protect (https://www.un.org/en/genocideprevention/about-responsibility-to-protect.shtml).

Waltz, K. (1959) *Man, the State and War: A Theoretical Analysis.* New York: Columbia University Press.

Wrangham, R. (2019) *The Goodness Paradox: The Strange Relationship between Virtue and Violence in Human Evolution.* New York: Vintage Books.

Chapter 6 Groups: Nations, states, sovereignty and immigration

Anderson, B. (1996) *Imagined Communities: Reflections on the Origin and Spread of Nationalism.* London and New York: Verso.

Conway, D. (2004) *In Defence of the Realm: The Place of Nations in Classical Liberalism.* Aldershot: Ashgate.

Friedman, M. (1978) What is America? In *The Economics of Freedom.* Cleveland: Standard Oil Company of Ohio.

Gellner, E. (1983) *Nations and Nationalism.* Oxford: Blackwell.

Hayek, F. A. (1978) Integrating immigrants. *The Times,* 9 March.

Hume, D. (1932 [1727–76]) *The Letters of David Hume* (two vols) (ed. J. Y. T. Greig). Oxford University Press.

Hume, D. (1985 [1777]) *Essays: Moral, Political and Literary* (ed. E. F. Miller). Indianapolis: Liberty Fund.

Jackson, R. H. (2007) *Sovereignty: Evolution of an Idea.* Cambridge: Polity Press.

Kukathas, C. (2021) *Immigration and Freedom.* Princeton University Press.

Kedourie, E. (1993) *Nationalism.* Oxford: Blackwell.

Lal, D. (2004) *In Praise of Empires: Globalization and Order.* Basingstoke and New York: Palgrave Macmillan.

Lomasky, L. E. and Tesón, F. R. (2015) *Justice at a Distance: Extending Freedom Globally.* Cambridge University Press.

Smith, A. D. (1999) *Myths and Memories of the Nation.* Oxford University Press.

Somin, I. (2020) *Free to Move: Foot Voting, Migration and Political Freedom.* Oxford University Press.

Van der Vossen, B. and Brennan, J. (2018) *In Defence of Openness: Why Global Freedom Is the Humane Solution to Global Poverty.* Oxford University Press.

Watson, A. (1982) *Diplomacy: The Dialogue between States.* London: Routledge.

Ypi, L. and Fine, S. (2016) *Migration in Political Theory: The Ethics of Movement and Membership.* Oxford University Press.

Chapter 7   Violence: Balance of power, war, military intervention

Booth, K. and Wheeler, N. J. (2008) *The Security Dilemma: Fear, Cooperation and Trust in World Politics.* Basingstoke and New York: Palgrave Macmillan.

Bull, H. (1995) *The Anarchical Society: A Study of Order in World Politics.* Basingstoke and London: Macmillan.

Friedman, M. (1962) *Capitalism and Freedom.* University of Chicago Press.

Higgs, R. (2004) *Against Leviathan: Government Power and a Free Society.* Oakland: The Independent Institute.

Higgs, R. (2005) *Resurgence of the Warfare State: The Crisis Since 9/11.* Oakland: The Independent Institute.

Hume, D. (1987) *Essays: Moral, Political, and Literary.* Indianapolis: Liberty Fund.

Little, R. (2007) *The Balance of Power in International Relations: Metaphors, Myths and Models.* Cambridge University Press.

Reed, C. and Ryall, D. (2007) *The Price of Peace: Just War in the Twenty-First Century.* Cambridge University Press.

Rengger, N. (2013) *Just War and International Order: The Uncivil Conditions in World Politics.* Cambridge University Press.

Rengger, N. (2017) *The Anti-Pelagian Imagination in Political Theory and International Relations.* London and New York: Routledge.

Sobek, D. (2009) *The Causes of War.* Cambridge: Polity.

Suganami, H. (1996) *On the Causes of War.* Oxford: Clarendon Press.

United Nations (2021) Responsibility to protect (https://www.un .org/en/genocideprevention/about-responsibility-to-protect .shtml).

Van de Haar, E. R. (2011) Hayekian spontaneous order and the international balance of power. *Independent Review* 16: 101–18.

Van de Haar, E. R. (2013b) David Hume and Adam Smith on international ethics and humanitarian intervention. In *Just and Unjust Military Intervention: European Thinkers from Vitoria to Mill* (ed. S. Recchia and J. M. Welsh), pp. 154–75. Cambridge University Press.

Walzer, M. (1992) *Just and Unjust Wars: A Moral Argument with Historical Illustrations.* New York: Basic Books.

Walzer, M. (2005) *Arguing about War.* New Haven and London: Yale University Press.

Chapter 8   Rules: International law and international organisation

Besson, S. and Tasioulas, J. (2010) *The Philosophy of International Law.* Oxford University Press.

Equality and Human Rights Commission (2017) What is the European Convention on Human Rights? (https://www.equal ityhumanrights.com/en/what-european-convention-human -rights).

Mises, L. von (1985) *Liberalism: The Classical Tradition.* Irvington-on-Hudson: Foundation for Economic Education.

Nabulsi, K. (2005) *Traditions of War: Occupation, Resistance, and the Law.* Oxford University Press.

Simmons, B. A. and Steinberg, R. H. (2006) *International Law and International Relations.* Cambridge: IO Foundation / Cambridge University Press.

Tuck, R. (1999) *The Rights of War and Peace: Political Thought and the International Order – From Grotius to Kant.* Oxford University Press.

Chapter 9   Economics: Trade, globalisation and development aid

Banerjee, A. V. and Duflo, E. (2011) *Poor Economics: Barefoot Hedge-fund Managers, DIY Doctors and the Surprising Truth about Life on Less than 1$ a Day.* London: Penguin.

Bauer, P. T. (1971) *Dissent on Development: Studies and Debates in Development Economics.* London: Weidenfeld and Nicolson.

Bauer, P. T. and Yamey, B. S. (1957) *The Economics of Under-Developed Countries.* Cambridge University Press.

Bhagwati, J. (2004) *In Defence of Globalization.* Oxford University Press.

Butler, E. (2021) *Introduction to Trade and Globalisation.* London: Institute of Economic Affairs.

Coker, C. (2014) *Can War Be Eliminated?* Cambridge: Polity.

Easterly, W. (2002) *The Elusive Quest for Growth: Economist's Adventures and Misadventures in the Tropics.* Cambridge, MA and London: The MIT Press.

Easterly, W. (2013) *The Tyranny of Experts: Economists, Dictators and the Forgotten Rights of the Poor.* New York Basic Books.

Germann, J. (2018) *Marxism.* In *International Organization and Global Governance* (ed. T. G. Weiss and R. Wilkinson), pp. 170–79. Routledge.

Hammarlund, P. A. (2005) *Liberal Internationalism and the Decline of the State: The Thought of Richard Cobden, David Mitrany and Kenichi Ohmae.* Basingstoke and New York: Palgrave Macmillan.

Hoekman, B. and Kostecki, M. (2009) *The Political Economy of the World Trading System: The WTO and Beyond.* Oxford University Press.

Hont, I. (2005) *Jealousy of Trade: International Competition and the Nation-State in Historical Perspective.* Cambridge and London: The Belknap Press of Harvard University Press.

Hume, D. (1987) *Essays: Moral, Political, and Literary.* Indianapolis: Liberty Fund.

Irwin, D. A. (1996) *Against the Tide: An Intellectual History of Free Trade.* Princeton University Press.

Irwin, D. A. (2017) *Clashing over Commerce: A History of US Trade Policy.* The University of Chicago Press.

Lal, D. (2002) *The Poverty of 'Development Economics'.* London: Institute of Economic Affairs.

List, F. (1841) *The National System of Political Economy* (trans. S. Lloyd). London: Longmans, Green and Co. (See The National System of Political Economy, Online Library of Liberty (libertyfund.org).

Moyo, D. (2009) *Dead Aid: Why Aid Is Not Working and How There Is Another Way for Africa.* London: Penguin.

Norberg, J. (2001) *In Defence of Global Capitalism.* Stockholm: Timbro.

Norberg, J. (2017) *Progress: Ten Reasons to Look Forward to the Future.* London: One World.

Palmer, T. and Warner, M. (2022) *Development with Dignity: Self-Determination, Localization and the End to Poverty.* New York: Routledge.

Panagariya, A. (2019) *Free Trade and Prosperity: How Opennesss Helps Developing Countries Grow Richer and Combat Poverty.* Oxford University Press.

Ricardo, D. (2002 [1817]) *The Principles of Political Economy and Taxation.* London: Empiricus Books.

Rosling, H. (with O. Rosling and A. Rossling Rönnlund) (2018) *Factfulness: Ten Reasons Why We Are Wrong about the World – and Why Things Are Better than You Think.* London: Sceptre.

Sally, R. (1998) *Classical Liberalism and International Economic Order: Studies in Theory and Intellectual History.* London: Routledge.

Sally, R. (2008) *Trade Policy, New Century: The WTO, FTAs and Asia Rising.* London: Institute of Economic Affairs.

Smith, A. (1981) *An Inquiry into the Nature and Causes of the Wealth of Nations.* Indianapolis: Liberty Fund.

Suganami, H. (1996) *On the Causes of War.* Oxford: Clarendon Press.

Van de Haar, E. R. (2010) The liberal divide over trade, war and peace. *International Relations* 24: 132–54.

Van de Haar, E. R. (2011) Philippine trade policy and the Japan–Philippines Economic Partnership Agreement (JPEPA). *Contemporary Southeast Asia* 33: 113–39.

Van de Haar, E. R. (2020a) Free trade does not foster peace. *Economic Affairs* 40: 281–86.

Van de Haar, E. R. (2020b) Rejoinder. *Economic Affairs* 40: 454–56.

Wolf, M. (2005) *Why Globalization Works.* New Haven and London: Yale University Press.

Chapter 10   Liberal IR theories

Ashworth, L. M. (1999) *Creating International Studies: Angell, Mitrany and the Liberal Tradition.* Aldershot: Ashgate.

Baldwin, D. A. (1993) Neoliberalism, neorealism, and world politics. In *Neorealism and Neoliberalism: The Contemporary Debate* (ed. D. A. Baldwin), pp. 3–25. New York: Columbia University Press.

Barbieri, K. (2005) *The Liberal Illusion: Does Trade Promote Peace?* Ann Arbor: The University of Michigan Press.

Bernstein, S. and Pauly, L. W. (2007) *Global Liberalism and Political Order: Toward a New Grand Compromise?* Albany: State University of New York Press.

Brown, M. E., Lynn-Jones, S. M. and Miller, S. E. (eds) (1996) *Debating the Democratic Peace.* Cambridge, MA: MIT Press.

Burchill, S. (2013) Liberalism. In *Theories of International Relations* (ed. S. Burchill and A. Linklater), pp. 57–87. Basingstoke and New York: Palgrave Macmillan.

Copeland, D. C. (2015) *Economic Interdependence and War.* Princeton University Press.

Doyle, M. W. (1986) Liberalism and world politics. *American Political Science Review* 80: 1151–69.

Doyle, M. W. (1996) Kant, liberal legacies and foreign affairs. In *Debating the Democratic Peace* (ed. M. E. Brown, S. M.

Lynn-Jones and S. E. Miller), pp. 3–57. Cambridge, MA: MIT Press.

Doyle, M. W. (1997) *Ways of War and Peace: Realism, Liberalism and Socialism*. New York and London: W. W. Norton & Company.

Dunne, T. (2005) Liberalism. In *The Globalization of World Politics: An Introduction to International Relations* (ed. J. Baylis and S. Smith), pp. 185–201. Oxford University Press

Gartzke, E. (2007) The capitalist peace. *American Journal of Political Science* 51(1): 166–91.

Goddard, S. E. and Krebs, R. R. (2021) Legitimating primacy after the Cold War: how liberal talk matters to US foreign policy. In *Before and After the Fall: World Politics and the End of the Cold War* (ed. N. P. Monteiro and F. Bartel), pp. 132–50. Cambridge University Press.

Griffiths, M. (2011) *Rethinking International Relations Theory*. New York and Basingstoke: Palgrave Macmillan.

Hammarlund, P. A. (2005) *Liberal Internationalism and the Decline of the State: The Thought of Richard Cobden, David Mitrany and Kenichi Ohmae*. Houndmills and New York: Palgrave Macmillan.

Jackson, R. H. and Sørensen, G. (2003) *Introduction to International Relations: Theories and Approaches*. Oxford University Press.

Jahn, B. (2013) *Liberal Internationalism: Theory, History, Practice*. Basingstoke: Palgrave Macmillan.

Jørgensen, K. E. (2018) *International Relations Theory: A New Introduction*. London: Palgrave.

Jørgensen, K. E. (2021) *The Liberal International Theory Tradition in Europe*. Basingstoke: Palgrave Macmillan.

Keohane, R. O. (1990) International liberalism reconsidered. In *The Economic Limits to Modern Politics* (ed. J. Dunn), pp. 165–94. Cambridge University Press.

Keohane, R. O. (2012) Twenty years of institutional liberalism. *International Relations* 26: 125–38.

Keohane, R. O. and Nye, J. N. (1989) *Power and Interdependence.* Cambridge, MA: HarperCollins.

Lang, A. F. (2015) *International Political Theory: An Introduction.* New York and Basingstoke: Palgrave.

MacMillan, J. (1998) *On Liberal Peace: Democracy, War, and the International Order.* London: I. B. Tauris Publishers.

Maersheimer, J. J. (2018) *The Great Delusion: Liberal Dreams and International Realities.* New Haven and London: Yale University Press.

Mansfield, E. D. (1994) *Power, Trade, and War.* Princeton University Press.

Mansfield, E. D. and Pollins, B. M. (eds) (2003) *Economic Interdependence and International Conflict: New Perspectives on an Enduring Debate.* Ann Arbor, MI: University of Michigan Press.

Mansfield, E. D. and Snyder, J. (2005) *Electing to Fight: Why Emerging Democracies Go to War.* Cambridge and London: MIT Press.

Martin, L. L. (2007) Neoliberalism. In *International Relations Theories. Discipline and Diversity* (ed. T. Dunne, M. Kurki and S. Smith), pp. 109–26 Oxford University Press.

McDonald, P. J. (2009) *The Invisible Hand of Peace: Capitalism, the War Machine and International Relations Theory.* Cambridge University Press.

Molloy, S. (2017) *Kant's International Relations: The Political Theology of Perpetual Peace.* Ann Arbor, MI: University of Michigan Press.

Nye, J. N. (1988) Neorealism and neoliberalism. *World Politics* XL: 235–51.

Owen, J. M. (1996) How liberalism produces democratic peace. In *Debating the Democratic Peace* (ed. M. E. Brown, S. M. Lynn-Jones and S. E. Miller), pp. 116–54. Cambridge, MA: MIT Press.

Owen, J. M. (1997) *Liberal Peace, Liberal War: American Politics and International Security.* Ithaca, NY: Cornell University Press.

Panke, D. and Risse, T. (2007) Liberalism. In *International Relations Theories: Discipline and Diversity* (ed. T. Dunne, M. Kurki and S. Smith), pp. 89–108 Oxford University Press.

Rawls, J. (1999) *A Theory of Justice*, revised edn. Oxford University Press.

Rawls, J. (2002) *The Law of Peoples, with 'The Idea of Public Reason Revisited'.* Cambridge and London: Harvard University Press.

Rengger, N. (2013) Realism tamed or liberalism betrayed? Dystopic liberalism and the international order. In *After Liberalism? The Future of Liberalism in International Relations* (ed. R. Friedman, K. Oskanian and R. Pacheco Pardo), pp. 51–66. London: Palgrave Macmillan.

Richardson, J. L. (2001) Contending liberalisms in world politics. ideology and power. Boulder, CO: Lynne Rienner.

Ruggie, J. G. (1982) International regimes, transactions, and change: embedded liberalism in the postwar economic order. *International Organization* 36: 379–415.

Schneider, G. and Gleditsch, P. (eds) (2013) *Assessing the Capitalist Peace.* London: Routledge.

Sinha, A. (2017) *John J. Mearsheimer on 'Liberal Ideals and International Realities'*. Yale MacMillan Centre, 30 November (https://macmillan.yale.edu/news/john-j-mearsheimer-liberal-ideals-and-international-realities).

Stein, A. R. (2008) Neoliberal institutionalism. In *The Oxford Handbook of International Relations* (ed. C. Reus-Smit and D. Snidal), pp. 201–21. Oxford University Press.

Van de Haar, E. R. (2021) Classical liberalism and IR theory. In *The Liberal International Theory Tradition in Europe* (ed. K. E. Jörgensen), pp. 119–32. New York and Basingstoke: Palgrave Macmillan.

Wilson, W. (1918) The fourteen points. National WWI Museum and Memorial (theworldwar.org).

Young, O. R. (1989) *International Cooperation: Building Regimes for Natural Resources and the Environment*. Ithaca and London: Cornell University Press.

Zacher, M. W and Matthew, R. A. (1995) Liberal international theory: common threads, divergent strands. In *Controversies in International Relations Theory: Realism and the Neoliberal Challenge* (ed. C. W. Kegley Jr), pp. 107–50. New York: St. Martin's Press.

Chapter 11 Classical liberalism, IR theory and foreign policy

Bull, H. (1966) The Grotian conception of international society. In *Diplomatic Investigations. Essays in the Theory of International Politics* (ed. H. Butterfield and M. Wight), pp. 51–73. London: George Allen & Unwin.

Bull, H. (1977) *The Anarchical Society: A Study of Order in World Politics*. Basingstoke and London: Macmillan.

Bull, H. (1990) The importance of Grotius in the study of international relations. In *Hugo Grotius and International Relations* (ed. H. Bull, B. Kingsbury and A. Roberts), pp. 65–94. Oxford: Clarendon Press.

Buzan, B. (2004) *From International to World Society? English School Theory and the Social Structure of Globalisation.* Cambridge University Press.

Buzan, B. (2014) *An Introduction to the English School of International Relations: A Societal Approach.* Cambridge: Polity Press.

Hill, C. (2003) *The Changing Politics of Foreign Policy.* Basingstoke and New York: Palgrave Macmillan.

Hume, D. (1998) *An Enquiry concerning the Principles of Morals.* Oxford University Press.

Linklater, A. and Suganami, H. (2006) *The English School of International Relations: A Contemporary Reassessment.* Cambridge University Press.

Van de Haar, E. R. (2008) David Hume and international political theory: a reappraisal. *Review of International Studies* 34(2): 225–42.

Van de Haar, E. R. (2021) Classical liberalism and IR theory. In *The Liberal International Theory Tradition in Europe* (ed. K. E. Jörgensen), pp. 119–32. New York and Basingstoke: Palgrave Macmillan.

Waltz, K. N. (2010) *Theory of International Politics.* Longgrove: Waveland Press.

Whelan, F. G. (2004) *Hume and Machiavelli: Political Realism and Liberal Thought.* Lanham and Oxford: Lexington.

Wight, M. (1991) *International Theory: The Three Traditions* (ed. G. Wight and B. Porter). London: Leicester University Press for the Royal Institute of International Affairs.

Chapter 12 Libertarians and IR

Block, W. (2003) National defense and the theory of externalities, public goods, and clubs. In *The Myth of National Defense: Essays in the Theory and History of Security Production* (ed. H.-H. Hoppe), pp. 301–35. Auburn: Ludwig von Mises Institute.

Bresiger, G. (1997) Laissez faire and Little Englanderism: the rise, fall, rise and fall of the Manchester School. *Journal of Libertarian Studies* 13: 45–79.

Carpenter, T. G. (ed.) (1989) *Collective Defense or Strategic Independence? Alternative Strategies for the Future.* Washington, DC, and Lanham: Cato Institute and Lexington Books.

Carpenter, T. G. (1995) Toward strategic independence: protecting vital American interests. *Brown Journal of World Affairs* 2(2): 7–13.

Carpenter, T. G. (2002) *Peace and Freedom: Foreign Policy for a Constitutional Republic.* Washington, DC: Cato Institute.

Cobden, R. (1878) *The Political Writings of Richard Cobden.* London: William Ridgway.

De Molinari, G. (1849) The production of security. The Mises Institute.

Denson, J. V. (2003) War and American freedom. In *The Costs of War: America's Pyrrhic Victories* (ed. J. V. Denson). New Brunswick and London: Transaction Publishers.

Ebeling, R. M. (2003) Classical liberalism in the twenty-first century: war and peace. In *Liberty, Security, and the War on*

Terrorism (ed. R. M. Ebeling and J. G. Hornberger). Fairfax: The Future of Freedom Foundation.

Ebeling, R. M. and Hornberger, J. G. (eds) (1995) *The Case for Free Trade and Open Immigration*. Fairfax: The Future of Freedom Foundation.

Ebeling, R. M. and Hornberger, J. G. (eds) (1996) *The Failure of America's Foreign Wars*. Fairfax: The Future of Freedom Foundation.

Ebeling, R. M. and Hornberger, J. G. (eds) (2003) *Liberty, Security, and the War on Terrorism*. Fairfax: The Future of Freedom Foundation.

Eland, I. (2004) *The Empire Has No Clothes: U.S. Foreign Policy Exposed*. Oakland: The Independent Institute.

Francis, M. (2007) *Herbert Spencer and the Invention of Modern Life*. Ithaca, NY: Cornell University Press.

Gotthelf, A. and Salmieri, G. (eds) (2016) *A Companion to Ayn Rand*. Chichester: Wiley Blackwell.

Grammp, W. D. (1960) *The Manchester School of Economics*. Stanford and London: Stanford University Press and Oxford University Press.

Gray, T. S. (1996) *The Political Philosophy of Herbert Spencer: Individualism and Organicism*. Aldershot: Avebury.

Hammarlund, P. (2005), *Liberal Internationalism and the Decline of the State: The Thought of Richard Cobden, David Mitrany, and Kenichi Ohmae*. Basingstoke: Palgrave Macmillan.

Higgs, R. (2005) *Resurgence of the Warfare State: The Crisis Since 9/11*. Oakland: The Independent Institute.

Higgs, R. (2006) *Depression, War, and Cold War: Studies in Political Economy*. Oxford University Press.

Hirst, F. W. (ed.) (1968) *Free Trade and Other Fundamental Doctrines of the Manchester School.* New York: Augustus M. Kelley Publishers.

Hoppe, H.-H. (2003a). Government and the private production of defense. In *The Myth of National Defense: Essays in the Theory and History of Security Production* (ed. H.-H. Hoppe). Auburn: Ludwig von Mises Institute.

Hoppe, H.-H. (ed.) (2003b) *The Myth of National Defense: Essays on the Theory and History of Security Production.* Auburn: Ludwig von Mises institute.

Howe, A. and Morgan, S. (eds) (2006) *Rethinking Nineteenth-Century Liberalism: Richard Cobden Bicentenary Essays.* Aldershot: Ashgate.

Journo, E. (ed.) (2009) *Winning the Unwinnable War: America's Self-Crippled Response to Islamic Totalitarianism.* Lanham: Lexington Books.

Lomasky, L. E. and Tesón, F. R. (2015) *Justice at a Distance: Extending Freedom Globally.* Cambridge University Press.

Prebble, C. A. (2019) *Peace, War, and Liberty: Understanding US Foreign Policy.* Washington, DC: Libertarianism.org.

Rand, A. (1964) *The Virtue of Selfishness: A New Concept of Egoism.* New York: Meridian.

Rockwell, L. H. (2003) *Speaking of Liberty.* Auburn: Ludwig von Mises Institute.

Rothbard, M. N. (2000) *Egalitarianism as a Revolt against Nature, and Other Essays.* Auburn: Ludwig von Mises Institute.

Rothbard, M. N. (2002) *The Ethics of Liberty.* New York and London: New York University Press.

Rothbard, M. N. (2003) America's two just wars: 1775 and 1861. In *The Costs of War: America's Pyrrhic Victories* (ed.

J. V. Denson). New Brunswick and London: Transaction Publishers.

Rothbard, M. N. (2004) *Man, Economy, and State: A Treatise on Economic Principles, with Power and Market – Government and the Economy.* Auburn: Ludwig von Mises Institute.

Spencer, H. (1978) *The Principles of Ethics* (two vols). Indianapolis: Liberty Fund.

Spencer, H. (1982) *The Man versus the State: With Six Essays on Government, Society and Freedom.* Indianapolis: Liberty Fund.

Trentmann, F. (2006) The resurrection and decomposition of Cobden in Britain and the West: an essay in the politics of reputation. In *Rethinking Nineteenth-Century Liberalism: Richard Cobden Bicentenary Essays* (ed. A. Howe and S. Morgan). Aldershot: Ashgate.

Van de Haar, E. R. (2010) The liberal divide over trade, peace and war. *International Relations* 24(2): 132–54.

Van De Haar, E. R. (2019) Fostering liberty in international relations theory: the case of Ayn Rand. *International Politics* 56: 1–16.

Van De Haar, E. R. (2020a) Free trade does not foster peace. *Economic Affairs* 40: 281–86.

Weinstein, D. (1998) *Equal Freedom and Utility: Herbert Spencer's Liberal Utilitarianism.* Cambridge University Press.

ABOUT THE IEA

The Institute is a research and educational charity (No. CC 235 351), limited by guarantee. Its mission is to improve understanding of the fundamental institutions of a free society by analysing and expounding the role of markets in solving economic and social problems.

The IEA achieves its mission by:

- a high-quality publishing programme
- conferences, seminars, lectures and other events
- outreach to school and college students
- brokering media introductions and appearances

The IEA, which was established in 1955 by the late Sir Antony Fisher, is an educational charity, not a political organisation. It is independent of any political party or group and does not carry on activities intended to affect support for any political party or candidate in any election or referendum, or at any other time. It is financed by sales of publications, conference fees and voluntary donations.

In addition to its main series of publications, the IEA also publishes (jointly with the University of Buckingham), *Economic Affairs*.

The IEA is aided in its work by a distinguished international Academic Advisory Council and an eminent panel of Honorary Fellows. Together with other academics, they review prospective IEA publications, their comments being passed on anonymously to authors. All IEA papers are therefore subject to the same rigorous independent refereeing process as used by leading academic journals.

IEA publications enjoy widespread classroom use and course adoptions in schools and universities. They are also sold throughout the world and often translated/reprinted.

Since 1974 the IEA has helped to create a worldwide network of 100 similar institutions in over 70 countries. They are all independent but share the IEA's mission.

Views expressed in the IEA's publications are those of the authors, not those of the Institute (which has no corporate view), its Managing Trustees, Academic Advisory Council members or senior staff.

Members of the Institute's Academic Advisory Council, Honorary Fellows, Trustees and Staff are listed on the following page.

The Institute gratefully acknowledges financial support for its publications programme and other work from a generous benefaction by the late Professor Ronald Coase.

Other IEA publications

Comprehensive information on other publications and the wider work of the IEA can be found at www.iea.org.uk. To order any publication please see below.

Personal customers

Orders from personal customers should be directed to the IEA:

IEA
2 Lord North Street
FREEPOST LON10168
London SW1P 3YZ
Tel: 020 7799 8911, Fax: 020 7799 2137
Email: sales@iea.org.uk

Trade customers

All orders from the book trade should be directed to the IEA's distributor:

Ingram Publisher Services UK
1 Deltic Avenue
Rooksley
Milton Keynes MK13 8LD
Tel: 01752 202301, Fax: 01752 202333
Email: ipsuk.orders@ingramcontent.com

IEA subscriptions

The IEA also offers a subscription service to its publications. For a single annual payment (currently £42.00 in the UK), subscribers receive every monograph the IEA publishes. For more information please contact:

Subscriptions
IEA
2 Lord North Street
FREEPOST LON10168
London SW1P 3YZ
Tel: 020 7799 8911, Fax: 020 7799 2137
Email: accounts@iea.org.uk